Denis Noble, CBE, FRS, is Emeritus Professor of Cardiovascular Physiology at the University of Oxford. He was Chairman of the IUPS (International Union of Physiological Sciences) World Congress in 1993, and Secretary-General of IUPS from 1993–2001. His previous publications include the seminal set of essays *The Logic of Life* (Boyd and Noble, OUP 1993), and he played a major role in launching the Physiome Project, one of the international components of the systems biology approach. *Science* magazine included him amongst its review authors for its issue devoted to the subject in 2002.

Non sai tu che la nostra anima è composta di armonia?
[Do you not know that our soul is composed of harmony?]

Leonardo da Vinci, *Trattato della Pittura*

The MUSIC *of* LIFE

Biology beyond Genes

Denis Noble

OXFORD
UNIVERSITY PRESS

OXFORD
UNIVERSITY PRESS

Great Clarendon Street, Oxford OX2 6DP

Oxford University Press is a department of the University of Oxford.
It furthers the University's objective of excellence in research, scholarship,
and education by publishing worldwide in

Oxford New York

Auckland Cape Town Dar es Salaam Hong Kong Karachi
Kuala Lumpur Madrid Melbourne Mexico City Nairobi
New Delhi Shanghai Taipei Toronto

With offices in

Argentina Austria Brazil Chile Czech Republic France Greece
Guatemala Hungary Italy Japan Poland Portugal Singapore
South Korea Switzerland Thailand Turkey Ukraine Vietnam

Oxford is a registered trade mark of Oxford University Press
in the UK and in certain other countries

Published in the United States
by Oxford University Press Inc., New York

British Library Cataloguing in Publication Data

Data available

Typeset by SPI Publisher Services, Pondicherry, India
Printed and bound in Great Britain by Clays Ltd, Elcograf S.p.A.

ISBN 978-0-19-922836-2

13

Contents

Introduction

'What is life?' The question can be interpreted in many ways. One way to approach it is scientifically. Even from this standpoint there can be a variety of answers since contemporary scientists can understand the question rather differently. Moreover, each generation needs to revisit the question almost anew – the advances in biological science are that rapid.

It is only 50 years ago that we first discovered that the genetic material was the molecule called DNA (deoxyribonucleic acid), and that it came in long molecular strands of four similar chemicals, called bases. Now:

- We know that the human genome, the entire DNA of a human, is a sequence of 3 billion pairs of bases – and we have identified each of them.
- We also know how the organisation of these bases serves to enable protein production. For each protein, the genetic material provides something like a template. The structural sequence of the proteins is encoded in the DNA. We know in some detail how this code works.
- For that matter, we also know the sequences and structure of many of the proteins that the DNA encodes.

Biological science has never advanced so rapidly.

How has that changed the way we see life? It has answered many questions and thrown up many more. The answers that we arrive at reflect the process of investigation that we follow. Over the last half century, we have proceeded by breaking living systems down into their smallest components, the individual genes and molecules. Humpty-Dumpty has been smashed into billions of fragments. This is an impressive achievement.

For example, we can now pinpoint a gene mutation whose effects may 'kick in' during middle age to cause sudden cardiac death. We know nearly all the major steps in this causal chain, though not yet why it kicks in precisely when it does in a given individual. This kind of success is more and more common. Yet, such examples are not appearing with the frequency that optimists predicted when the human genome project was announced. The benefits for healthcare are slow to arrive.

Why is that? People are beginning to understand the reason. It has to do with how the small scale relates to the large. We know a lot about molecular mechanisms. Now the challenge is to extend that knowledge up the scale. How do we use it to throw light on the processes that govern entire living systems? That is not an easy question. Quite soon, as we move from genes to the proteins that they code for, and then on to the interactions between these proteins, the problems become seriously complicated. Yet we need to understand these complexities in order to interpret the molecular and genetic data, and on that basis to talk in a fresh and useful way about larger questions like 'What is life?'

This, then, is the challenge that sequencing the genome has raised. Can we put Humpty-Dumpty back together again? That is where 'systems biology' comes in. This is a new and important dimension of biological science, though it has strong historical roots in classical biology and physiology going back over a century. In recent decades, however, biologists have tended to focus quite narrowly on the individual components of living organisms. What properties does each component have? How does it therefore interact, over the short term, with other components of similar scale? Now, we are ready to ask some bigger questions. These are about systems. At each level of the organism, its various components are embedded in an integrated network or system. Each such system has its own logic. It is not possible to understand that logic merely by investigating the properties of the system's components.

This book is about systems biology. It is also about the preconditions for, and implications of systems biology. It says that at this stage in our exploration of life, we need to be ready for a basic re-think.

Molecular biology requires a certain way of thinking. It is about the naming and behaviour of the parts. We reduce each whole to its component parts and define them exhaustively. Biologists are now perfectly used to that thinking and the interested lay public has caught up, too. So we are now ready to move on. Systems biology is where we are moving to. Only, it requires a quite different mind-set. It is about putting together rather than taking apart, integration rather than reduction. It starts with what we have learned from the reductionist approach; and then it goes further. It requires that we develop ways of thinking about integration that are as rigorous as our reductionist procedures, but different. This is a major change. It has implications beyond the purely scientific. It means changing our philosophy, in the full sense of the term.

How to provoke such a change? I have chosen to write a polemic. This book is a radical analysis of many of the currently accepted dogmas in biology. It turns some of them upside down. It offers an unashamed defence of the need for a systems level approach. That is not because I am unimpressed with what reductionist molecular biology has achieved. On the contrary, it is because I want to see biological science garner the fruits that the great reductionist drive has put within our grasp.

As I explain in Chapter 5, I started my research career in physiology as a full 'card-carrying' reductionist. I know how successful reductionist science is done and have done much of it myself in my own field. I still use its methods quantitatively in my current research on simulating the organs of the body. And that is how, during the last decade or so, I have come to see the need to redress the balance. If we all keep our noses down to the lower-level grindstone, no-one will see the bigger picture, or realise what is needed if we are to fill it in. Successful integration at the systems level must be built on successful reduction, but reduction alone is far from sufficient.

Like any polemicist, I make free use of metaphor. I also tell some stories. These are intended to be enjoyable – and also to jolt the reader away from many current dogmas.

In 1944, Erwin Schrödinger wrote a remarkable book (Schrödinger 1944). In it, he correctly predicted that the genetic code is an 'aperiodic crystal', that is, a chemical sequence without regular repetition. Like many scientists at that time, he thought that the code would be found in the proteins rather than in DNA, so what he spoke of was not where he expected it – but it was there nonetheless. Many of his insights match remarkably well with what we have since learned. In just under 100 pages, he shifted the basic paradigms of biology.

This book is of similar length. I first thought to give it the same title: 'What is life?' But I have not been so audacious. Instead, I have chosen a title that reflects the main metaphor of the book: the systems-level view of life can be compared to music. If so, where is the score and who was the composer? A central question, therefore, that recurs throughout the book is 'Where, if anywhere, is the program of life?' The French Nobel Prize-winners Jacques Monod and François Jacob (Monod and Jacob 1961; Jacob 1970), referred to the 'genetic program' (le programme génétique): the idea that the instructions for the development of each living organism lie in its genes. The same idea is conveyed by the popular description of the genome as the 'book of life', a kind of blueprint. The central role of genes as causal agents was also greatly reinforced by popular perceptions of Richard Dawkins' highly influential book *The selfish gene* (Dawkins 1976).

The theme of my book is that there is no such program and that there is no privileged level of causality in biological systems. Chapter 1 lays the groundwork for the rest of the book. It does this first by recasting the genome as a database for the transmission of successful organisms, rather than a program that 'creates' them. The second step is to replace the metaphor of the 'selfish gene' by 'genes as prisoners'. These two radical switches of perception are essential to understanding the rest of the book. While it is necessary to deal with the popular (mis)perceptions of 'genetic programs', 'the book of life', and 'selfish genes', I hasten to acknowledge that the scientists responsible for these fruitful ideas may well not have approved of the way they have been widely interpreted. Richard Dawkins, for example, has also written some of the best critiques of the 'program' idea, and is himself far from being a gene determinist.

The book is organised into ten chapters. Each uses a different musical metaphor for some aspect of the biology of life. We start with the genome in Chapter 1 and end with the brain in Chapter 9. Chapter 10 stands on its own as a kind of coda.

Acknowledgements

I acknowledge valuable discussions on various aspects of this book with Geoff Bamford, Patrick Bateson, Steven Bergman, Sydney Brenner, Jonathan Cottrell, Christoph Denoth, Dario DiFrancesco, Yung Earm, David Gavaghan, Peter Hacker, Jonathan Hill, Peter Hunter, Otto Hutter, Roger Kayes, Anthony Kenny, Sung-Hee Kim, Junko Kimura, Peter Kohl, Jean-Jacques Kupiec, Ming Lei, Nicholas Leonard, Jie Liu, Denis Loiselle, Latha Menon, Alan Montefiore, Penny Noble, Ray Noble, Susan Noble, Carlos Ojeda, Etienne Roux, Ruth Schachter, Pierre Sonigo, Christine Standing, Richard Vaughan-Jones and Michael Yudkin. Many of these, and the Oxford University Press readers, also criticised early versions of various chapters. I benefited greatly from their feedback, though I am of course responsible for the errors and misunderstandings that remain.

I thank many friends and colleagues in East Asia who introduced me to aspects of their cultures that find expression in Chapters 8 and 10.

An early version of the Silman story in Chapter 1 appeared in French under the title 'Pourquoi il nous faut une théorie biologique' in the online publication *Vivant* in 2004. Parts of Chapter 3 are based on 'Is the genome the book of life?', *Physiology News* (2002), **46**, 18–20. The dialogue in Chapter 9 is based on 'Qualia and private languages', *Physiology News* (2004), **55**, 32–3, while the following story first appeared as 'Biological explanation and intentional behaviour' in *Modelling the mind* (ed. K.A. Mohyeldin Said *et al.*, Clarendon Press, Oxford, 1990), pp. 97–112. Some of the philosophical background was developed in *Goals, no goals and own goals* (ed. A. Montefiore and D. Noble, Unwin Hyman, London, 1989) and in various Novartis Foundation meetings on the nature of biological science.

1 § The CD of Life: the Genome

> They [genes] are all in the same boat.
> Maynard Smith and Szathmáry 1999

For humans at least, to live is to experience. How can we understand this?

One thing is clear: experience is grounded in matter. The connection is there for us to draw. But drawing it is quite a complicated task. And, alas, complexity is uncomfortable, so we are inclined to ignore it.

That tends to happen, for instance, when we try to tease out the linkage between human experience and material reality. We say, 'It's pretty simple, really'. But it isn't.

Introducing the Silmans

Consider an example. Before writing this page, I relaxed by listening for the first time for a long time to one of my favourite pieces of music: the piano trio in E-flat major by Schubert. I put the CD into my player and lay down on the sofa. As the music entered the slow movement, I cried.

The emotional effect of this piece of music, which I first heard live in a concert, is always very strong. We must all have our favourite pieces that have this effect on us. The effect does not always depend on the music itself. It can also depend on the context, the people we were with, and the significance of the event in our lives.

So, what caused me to cry?

Imagine some space travellers watching this scene. They are creatures from a world in which silicon replaces carbon. So let's call them Silmans. They have some of the characteristics associated in science fiction with 'androids'. They notice the crying. They record the sound waves in the room. As scientists, they trace the sequence of cause and effect, back through the loudspeakers, the amplifiers, the laser disc reader, right down to the CD itself.

One of them does a Silman version of 'Eureka!' 'I've found it', he says, as he explains to his colleagues that the whole effect is caused by some highly specific digital information on the CD. Another of the Silmans is nevertheless sceptical. 'How', he says, 'could just a bunch of numbers have this effect?'

The discoverer counters the scepticism by pointing out that this is the lowest level of the chain of cause and effect. Without the digital information, there would be no music, no emotion. Moreover, if you play around with that information, 'mutate' it as it were, by playing it too fast or too slow, or playing it backwards, transposing sections, or even transposing bits from another CD, then the person in the room no longer cries. In fact he may angrily turn the machine off and even throw the disc away.

There is an inevitable and mechanical chain of cause and effect here. Any experiment the Silmans might do would reinforce the one-way nature of this chain. Different amplifiers, speakers, and other gadgetry can replace everything except the highly specific digital information on the CD. Surely, then, they conclude that this is the cause of me crying.

Of course, we know better. We would say that the causes of my crying include:

- Schubert, because he wrote the music;
- the piano trio, because they played it with such heart-tugging inspiration;
- and the beautiful context in which I first heard the music and first cried as a result of it. This, we would say, is in my memory and forms the emotional context.

We would say that the digital information on the CD is just a way of capturing the moment, as accurately as possible, and making it possible for me to recreate, partially at least, the original moment. We know also that the information could be coded in many different ways, including analogue encoding in the form of a vinyl disc. It is just a database that enables the music to be stored and recreated.

In short, we would have no difficulty at all in laughing at the stupidity of our Silman visitors from another planet. They saw a simple explanation, we would say, and grabbed at it. How stupid! Well, we should be careful whom we laugh at. For we, too, get trapped in simplistic explanations.

DNA-mania

Indeed, there is a popular dogma that is reinforced daily in the media—and, it must be said, by many scientists—that rests on a crude mistake, just like the Silmans'. André Pichot[1] has called this DNA-mania. It is the delusion that the DNA code 'causes' life in much the same way as the CD 'caused' my experience of the Schubert piano trio.

The analogy is obvious. The human genome is in some ways a little like a CD. It carries digital information. Let's quickly summarise how. The genome is all the chromosomes in a cell. A chromosome is a long DNA molecule and some associated proteins. It is conventionally divided into genes. A gene is a section of DNA that is used in producing a particular protein.

DNA is composed of four chemicals (nucleotides), generally referred to by the letters A, T, G, and C.[2] There are two strands of DNA in each chromosome, wrapped around each other in a double helix. It was the discovery of this double helical structure that formed Watson and Crick's Nobel Prize-winning work in 1953. The nucleotides in one strand always lie opposite those in the other according to the rule: A goes opposite T, G opposite C. Two such

---

[1] A French philosopher and historian of science, author of *Histoire de la notion de gène* (Pichot 1999).

[2] Adenine, thymine, guanine, and cytosine.

complementary nucleotides make a base pair. The genome is 3 billion base pairs long. These form 20 000–30 000 genes.

In each gene, the chemicals are arranged in specific ways to facilitate the production of specific proteins. Every time a protein is needed, the appropriate chemical 'code' is 'read off' the gene; this gives the pattern of chemical elements that will make that protein what it is. Our genes encode the sequences of the 100 000 or so proteins that make up the human body. No protein is made that is not coded for by a gene. So the genome *is* important. After all, proteins are crucial for life.

A living cell is a continuing, action–packed drama. Molecules interact and change. One change triggers another, and so on and on. Complex chains of molecular interaction happen again and again. We call them 'pathways'. There are cell cycle pathways, which correspond to the cell 'ticking over'. There are developmental pathways, because cells grow, divide, and form more cells. There are all sorts of regulatory pathways. And proteins form the backbone of all these biochemical pathways.

Cells organise into tissues, such as skin, bone, muscle, to form organs such as the heart and kidneys, and finally, all these, together with the immune and hormonal systems, form the organism, the whole animal. This operates in many different ways, at various levels of organisation. And all of this 'function', as biologists say, involves proteins.

The causality seems to be entirely one-way. The DNA causes the proteins, the proteins cause the cells, and so on. The organism itself is just what shows on the outside; what is really happening, the inside story, is that the information coded in the genes is being expressed. In biologist-speak, the phenotype is 'created by' the genotype. The story is seductive.

We have fitted ourselves out with a magnificent set of blinkers. We have rendered ourselves incapable of looking at the relationships between the genetic code and living systems in any other way.

This chapter asks why.

- Why are we so fond of the gene-centred view? We can explore this question by examining a classic and very popular statement of this view—Dawkins' 1976 description of the 'selfish gene'.

- How did so many people come to interpret this view as genetic determinism? That question is particularly important since, as I will show, it is not that of Dawkins himself. Let us then explore the historical context, out of which DNA-mania has developed.

We start with the reductionist causal chain. This is the 'inside story' that we have just discussed. Schematically, it looks like this:

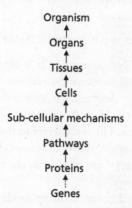

FIG 1. The reductionist causal chain.

The chain runs upwards. It is a 'one-way' system, from the genes to the organism. The idea is that, if we knew all about the lowest-level elements, genes and proteins, then everything about the organism would be clear to us. We could work out what happens at the higher levels, and explain it completely, in terms of our low-level knowledge. We could reconstruct the whole organism from the bottom up.

The first step in the chain is fainter than the others because it represents a rather different causal relationship. At each stage above this one, we are talking about physical causes—how one chemical reaction leads to another. But at the first stage something different is happening, over and above the physical causation of the chemical reactions involved. It is generally described as the reading of a code. There is transcription and translation of the code. This code is

sometimes called the blueprint of life, or the program of life, following Monod and Jacob's colourful idea of 'le programme génétique'
(Monod and Jacob 1961; Jacob 1970).

So much for the diagram. The problem with it is that it shows only
half the story. When we get to Chapter 4, we shall see how much it
misses out. But for the moment let us assume it is as comprehensive as
it is supposed to be.

On that assumption, then, let us ask: does the causal mechanism
work in the way that is represented here? By no means!

Problems with genetic determinism

Genes are coded as DNA sequences. It is these sequences that are
replicated and passed on to future generations. So biologists also call
genes replicators. Gene determinism somehow sees them as causal
agents. How can that be? After all, what does DNA *do*? As biological
molecules go, not much. The real players in the action of life are the
proteins. They are the really active molecules. They indulge most in
the biochemical processes necessary for life to occur. DNA is in
comparison rather passive.

Proteins are produced in tiny factories inside the cells of the body.
Biologists call them ribosomes. These factories get going when they
receive a message that 'tells' them to make a certain protein. Each
such message is generated using DNA. A DNA sequence that corresponds to the relevant protein sequence is copied onto another molecule, appropriately called a 'messenger', which transmits a form of
the sequence to the ribosomes. The messenger molecules, called messenger RNA (ribonucleic acid), are another kind of nucleic acid
sequence. The DNA sequences are therefore a kind of template, a
specific sequence of nucleotides that can be transcribed to produce
the message that is then translated into an amino-acid sequence when
the protein is made. (Amino acids are the units of which protein is
composed, just as nucleotides are the units of which DNA is
composed.)

That process is called 'gene expression'. This terminology gives the
impression that the whole process is implicit in the gene, or at least in

the information that the gene holds, which simply needs to be 'expressed'.

But it is a little odd to say, as we often do, that the DNA sequence 'determines' the protein. In fact, the DNA just sits there, and occasionally the cell reads off from it a sequence that it needs, in order to get some protein produced. This looks very much like my hi-fi equipment reading the digital information on a CD to generate the real 'action': the music. So the first step in the reductionist chain of cause and effect is not a simple causal event at all. When a sequence is read off, that is an important event, which initiates a whole series of subsequent events. These are physical events. True. But it is the process of reading that matters, as well as the object that is read.

This process involves certain systems of proteins. If we wish to identify an agent of the action, it must be those systems. They 'read' the DNA code. DNA does nothing outside the context of a cell[3] containing these protein systems, just as the CD can do nothing without the CD reader. So, we have the paradox that proteins are required for the machinery to read the code to produce the proteins. I will return to this paradox in later chapters.

But is this just a technicality? Whether we start the causal chain from genes or proteins doesn't matter much, perhaps. Don't we just have to adjust our story slightly and say that the genetic code lies in the protein sequences? That might be a reasonable way of looking at the matter, except that it assumes that each gene codes directly for a single protein, that is, that the two sequences, in the DNA and the protein, are straightforwardly identical. But they are not.

In higher animals, the bits of DNA code that we lump together and call collectively a 'gene' are not always continuous. In many, perhaps most cases, they are broken up into segments. These segments, called 'exons', are separated by non-coding stretches of DNA, called 'introns'. The exon codes can be combined in various orders to produce a full protein code. In ways we do not yet understand, the DNA threads are folded into a three-dimensional form in the nucleus of

---

[3] Viruses are not an exception to this rule. They need to enter a cell to use its machinery to reproduce. Outside a cell they cannot reproduce.

each cell. They cannot exist as a straight thread since each cell contains two metres of DNA, which is around 100 000 times longer than most cells. The way the threads are folded inside the cell may make the reading of certain sequences easier than others.

There can therefore be many different ways of reading the separate exons and joining them up. Technically, there are often many 'splice variants' of a gene, which can therefore code for a set of different proteins. These splice variants are the different ways of reading the separate exons and joining them up (Black 2000). Thus, if a gene consists of three exons, *a*, *b*, and *c*, it could be read as the forms *a*, *b*, *c*, *ab*, *bc*, *ac*, *abc*, and perhaps even as *cba*, *ca*, *ba*, each of which would code for a different protein. At present, we do not know the rules for which combinations are possible and used in coding for proteins.

Consider the gene called *Dscam* in the fruit fly *Drosophila*. It has 110 introns and therefore tens of thousands of possible splice variants (Celotto and Graveley 2001). Moreover the *Dscam* gene does not always operate in the same way. It changes its role with the life cycle of *Drosophila*. At any one stage, some of the theoretically possible splice variants will work and others will not. At earlier and later stages, the picture will be different.

To an extent, it depends on the cellular environment. For instance, there are proteins that affect the transcription of DNA sequences. Some activate transcription; some inhibit it; they interact in complex ways. At the same time, there are features in the DNA code itself that influence whether a particular variant can be expressed. Within the DNA sequences of a given gene we find promoter elements and enhancer sequences. So the regulation of gene expression, as we say, involves a multiplicity of factors which operate and inter-operate in subtle ways.

There is regulation of how the code is read off from the gene in order to form the protein (transcription), and regulation of what happens after transcription. These are all complicated processes subject to many other influences than the DNA code itself.

What this means is that there are many different ways to read a genome. So my analogy with a CD is limited. When you put a CD into your hi-fi, there is only one way you are going to get music out of each track. This process involves a single, one-way read-out. When it comes to genes, by contrast, we have flexible, combinatorial read-out. A clever CD reader can also do this to some extent. We can program our hi-fi equipment to play the music tracks in flexible orders depending on how the recording has been broken up to form the individual tracks. The difference is that the genome is broken up to an unimaginably greater degree, the consequences of which we will explore in Chapter 2.

Included in this flexibility are many back-up processes. So it is possible to correct for errors and failures at the genome level. Indeed, it can happen that an important gene is completely knocked out, and the organism still manages to get by. If Plan A does not work, Plan B clicks in—the cell is still able to form proteins which function in place of those the non-functioning gene was originally used for.

To these multiple influences at the bottom level of protein production we must add an important higher-level complexity. This is that there is no one-to-one correspondence between genes and biological functions. Strictly speaking, therefore, to speak of a gene as the 'gene for x' is *always* incorrect. Many gene products, the proteins, must act together to generate biological functions at a high level. If we must use the expression 'gene for x' then we should at least add the plural and speak of the 'genes for x'.

Even this way of speaking is, however, seriously misleading. Not only do many genes co-operate in coding for the proteins that inter-act to produce any given biological function, each gene may also play a role in many different functions, which makes it difficult to label genes with functions.

I am talking here about higher-level function in organisms. Let's think of some of these functions. The pacemaker rhythm of the heart would be one. Another would be the secretion of insulin by the pancreas. And then we could take the transmission of impulses in the brain. Then let's think of the lower-level biological processes involved in those functions. For instance there is the process whereby calcium ions get pushed out of cells. Certain identifiable proteins combine to produce this effect. They are important because calcium is used as a controller of many processes in cells and organs.

This process of moving calcium ions goes on in all sorts of ways and contexts. For instance, it is involved in all three of the functions we have identified, and many more besides. In fact it is hard to think of a single higher-level function that does not involve these calcium-pushing proteins—and so, implicitly, the genes that code for them. I could repeat the same story for many other processes in cells. Many of the lower-level processes are used again and again in many different functions. High-level functions are therefore like a game of recombinations.

Suppose we sat down to identify the role or roles that those genes play in high-level functions. We would end up with a list that went on almost forever. That is what happens when we start to study how biological function emerges. We get involved in an almost endless game of recombinations. So, while it is relatively easy to label genes with low-level functions, which proteins they code for, it is much more difficult to label genes with high-level functions.

We need a manual that lists all the functions a gene is involved in, and how it contributes to each. Nature does not provide one. We have to work this stuff out for ourselves. That is the research project we call gene ontology. And to get anywhere with that project, we have to look beyond genes and proteins. We need to study the higher-level functions.

This is my primary reason for opposing the otherwise colourful metaphor describing the genome as 'the book of life' (Chapter 3). A book may describe, explain, illustrate, and may do many other things, but if we opened it up to find just strings of numbers, like the machine code of a computer program, we would surely ask where is the book itself; we would say that we had only been given a database. We could, perhaps, use another, interpretative program to generate a 'book' from it; but, until then, all we would have would be a mass of ciphers.

My central argument will be that the book of life is life itself. It cannot be reduced to just one of its databases. For let's be clear that the genome is only *one* of the databases. Function in biological systems depends also on important properties of matter that are not specified by genes. We will return to this aspect in Chapter 3.

Origin of the appeal of genetic determinism

There is work to do before we can start to make much sense of the genetic information we have discovered. The problems are immense, as we shall see in Chapter 2. Indeed, we must wonder how long it will take us to overcome them.

Why, then, has the genetic determinist agenda had such a wide and fashionable appeal? How does it come to dominate the way in which public debate on genes takes place, with 'genes for this' and 'genes for that' appearing with regular frequency, implying that it is only a matter of time before we find the genes for everything? This is where we need to look at the history of the development of ideas on genetics and on biology as a science.

There is an interesting contrast between the ways in which these ideas have developed in French-speaking and English-speaking countries. I will refer to the debate in French-speaking countries later in this book. In the Anglo-Saxon world the debate has been dominated by arguments between the gene-centred views of people like Richard Dawkins (1976) and the multi-level selection views of people like Stephen Jay Gould (2002).

The gene-centred view, the 'selfish gene' view, is a metaphorical polemic: the invention of a colourful metaphor to interpret scientific discovery in a particular way. It has provided valuable insights and these have been used to advance biological science in novel ways. I am not one of those critics of 'the selfish gene' idea who deny its impact and value. But it is nevertheless a metaphor. It is not a straightforward empirical scientific hypothesis. To demonstrate this I want to challenge the reader to a thought experiment. I will first give you one of the central statements of the 'selfish gene' idea. I will then rewrite it so that each sub-phrase (except for one anodyne statement) is replaced by a possible alternative, founded on an opposing metaphor that will form the basis for the rest of this book. The challenge is to think of an empirical test that could possibly distinguish between these two diametrically opposed ways of seeing the relationship between genes and phenotypes.

First, then, the original 'selfish gene' statement (Dawkins 1976: 21):

Now they swarm in huge colonies, safe inside gigantic lumbering robots, sealed off from the outside world, communicating with it by tortuous indirect routes, manipulating it by remote control. They are in you and me; they created us, body and mind; and their preservation is the ultimate rationale for our existence.

I would like the reader to think carefully about this statement to absorb its full import. Ask yourself whether you find the statement self-evident, shocking, implausible, likely, true, false, nonsense. Is it theory, fact, or neither? Form a view about it before you continue. Whichever of these views you hold (and all have been expressed by readers of *The selfish gene*), I believe you will find the test an interesting challenge.

So, now let's see what happens when we replace each phrase, except for the phrase 'they are in you and me', by an alternative written from an opposing viewpoint, that of 'genes as prisoners':

Now they are trapped in huge colonies, locked inside highly intelligent beings, moulded by the outside world, communicating with it by complex processes, through which, blindly, as if by magic, function emerges. They are in you and me; we are the system that allows their code to be read; and their preservation is totally dependent on the joy we experience in reproducing ourselves. We are the ultimate rationale for their existence.

The experiment is made even more effective if we arrange the two statements in register:

Now they swarm in huge colonies, safe inside gigantic
Now they are trapped in huge colonies, locked inside highly
lumbering robots, sealed off from the outside world,
intelligent beings, moulded by the outside world,
communicating with it by tortuous indirect routes,
communicating with it by complex processes,
manipulating it by remote control.
through which blindly, as if by magic, function emerges.
They are in you and me;
They are in you and me;
they created us, body and mind;
we are the system that allows their code to be read;
and their preservation is the ultimate rationale for our existence.
and their preservation is totally dependent on the joy we experience in reproducing ourselves.
We are the ultimate rationale for their existence.

To many of my readers this test will appear strange and challenging. Such a different view of the same thing: surely scientists must already know which is correct? Yet, I have tried this test many times now, always with the same result, which is that no-one seems to be able to think of an experiment that would detect an empirical difference between the two statements. The statements cannot therefore be a matter of empirical science, except for the obviously correct statement 'they are in you and me', which is certainly empirical, but not a difference between the statements.

Dawkins and I agree entirely on this point. In a later book he writes 'I doubt that there is any experiment that could be done to prove my claim' (Dawkins 1982: 1). Dawkins also makes clear that he is far from being a genetic determinist. In *The selfish gene* he writes 'We have the power to defy the selfish genes of our birth' (p. 215). Even more clearly, a chapter of a more recent book (Dawkins 2003) is entitled 'Genes Aren't Us'. I find that readers of *The selfish gene* often ignore these aspects, so that his central 'selfish gene' statement is taken in isolation as a gene-determinist argument. At the end of this chapter I will explore the reasons why the central metaphor of *The selfish gene* has fuelled gene determinism.

The acid test of empirical scientific content in an argument is to see what happens when you try to unpack it by stating its opposite and ask for an empirical test between the two. If no such test exists, then we are dealing either with sociological, polemical viewpoints, which can differ according to the stance of the speaker, or we are dealing with metaphor, and, of course, we could be dealing with both since metaphor is a favourite recourse of polemicists. This book is also a polemic and, as such, it makes deliberate use of metaphor and metaphorical stories. Regarding genes as 'physiological prisoners' is just as much a metaphor as is describing them as 'selfish'. I do not claim any empirical basis for this alternative. Which of the descriptions you prefer is not primarily dependent on any scientific knowledge. There is good science to be colourfully interpreted by both of these metaphors. The problems lie in the limitations of those metaphors.

'Selfishness' is based on the idea that a gene that confers selective advantage on an organism thereby 'selfishly' ensures its own survival and transmission. By extension, a very limited form of altruism, that is, apparent 'non-selfishness', can arise through influencing the chances of gene survival in close relatives, an idea that was first propounded mathematically by William Hamilton. Such models can make sense, from the viewpoint of a 'selfish gene', of sacrificing oneself for the sake of the survival of one's own gene complement in relatives, and genes that favour this behaviour will improve their chances of survival. The gene-centred view looks at the individual interests of the gene in the competition for survival.

By contrast, the view of 'genes as prisoners' emphasises the need for individual genes to co-operate with many other genes to generate physiological function, so that they are strongly restricted by this requirement. This view highlights the fact that no gene is selected in isolation. Its survival depends on that of many other genes with which it co-operates in coding for the physiological functions that confer selective advantage. So genes that favour successful co-operation with other genes also improve their chances of survival.

Each metaphor has its advantages and disadvantages, and there may be no harm done in using them provided that we recognise what we are doing and that we correctly analyse how the metaphors work so that we know which parts of the metaphor source idea map well to the scientific target to which they are applied. At the end of this chapter I will explore the mapping of the competing metaphors for genes. Throughout this book I will also explore the limitations of my metaphors, including a dramatic ending in which the central metaphor of all simply disappears.

No metaphors map perfectly to the situation they are describing. They highlight certain aspects at the price of playing down others. The harm is done when we take the metaphors too literally, extend them beyond their range of application, and interpret them as uniquely correct scientifically. A useful antidote to this tendency in the case of reductionist metaphors applied to genes is to remember that genes, like the digital information on a CD, are 'only' bits of a database, that they do not deterministically 'program' anything on their own, and that it is organisms that live or die and so provide the basis for evolutionary selection. As you read this book, you will see why I take this view.

Life is not a protein soup

I have already referred to the fact that the debates on evolutionary theory in the Anglo-Saxon world have been largely dominated by that between Stephen J. Gould and Richard Dawkins. Although my position may reflect that of Gould rather more than that of Dawkins, it doesn't really fully fit either viewpoint. One of the problems is that the debate has been so polarized, and so finely balanced on meanings of words, that one sometimes has the impression more of a medieval debate than of a modern scientific one. A lot of air would be cleared immediately if the role of metaphor were to be openly recognised and analysed. This is rarely done in science, although there is an extensive literature on metaphor in studies of linguistics, philosophy, and cognitive psychology (Kövecses 2002; Lakoff and Johnson 2003).

Different, even competing, metaphors can illuminate different aspects of the same situation, each of which may be correct even though the metaphors themselves may be incompatible. We benefit most when we recognise that. We should therefore treat the competition between metaphors differently from that between descriptions that differ empirically. Metaphors compete for insight, and for criteria like simplicity, beauty, creativity, all of which we use in judging scientific theories over and above their empirical correctness. But ultimately it is by the empirical tests that scientific theories live and die.

We also need to recognise that much of what is stated about genes at the reductionist level comes close to circularity. Molecular success for a gene is reproducing itself as much as possible so that its frequency in the gene pool increases. At the 'selfish' gene-oriented level, we tend to ignore other criteria for success, such as integrative (collaborative) properties of a higher-level network which would apply to many genes simultaneously. Yet this is the biological reason for the success of a gene or, more correctly, a set of genes.

We need to see that the success of a gene lies in its involvement in the expression of high-level function. This, after all, is what enables a given organism to be favoured in the selection process. The logic for explaining the success of a gene therefore does not lie in its DNA code; it lies in how that code is interpreted and in how the results of that interpretation fit into the overall successful logic of life.

Fundamental questions like 'What is a gene?' need to be re-addressed. The answer is not so obvious when we have to take account of modular coding regions that form parts of many genes, of genes that code for many different proteins, and of genes that completely change their function as one species evolves from others. This raises the question whether a gene should be defined by its coding or by its function.

Furthermore, the three-dimensional arrangement of the DNA threads is important to how the DNA sequences are read. This may determine which splice variants are favoured over others. Nature has been opportunistic. It did not need to organise its genome database for the convenience of humans trying to read it. Nor did it start, like Adam naming the animals, by naming the functions each gene would serve. It has serendipitously explored possible functional combinations. Only a very tiny fraction of these actually make sense at a higher level. And it was at this higher level that success or failure was determined.

Looked at from the systems-level perspective I am advocating, genes and proteins are rather like the building blocks of a child's toy such as Lego. They are elements that can be arranged in many different ways, with the same elements playing different roles depending on how they are arranged to interact with other elements. That is why I lay stress on gene expression patterns. The same genes expressed in a different pattern can produce a quite different physiological function.

This, then, is the great challenge of twenty-first century biology: how to account for the phenotype in terms of the systems-level interactions of the proteins. About this, molecular genetics tells us very little. In fact, it relies on having this worked out in order properly to annotate the genome. We also need systems-level analysis to understand the feedback onto gene expression (Chapter 4). The genome needs to be read through the phenotype, not the other way round.

We have become transfixed by the great success in explaining protein sequences in terms of encoded DNA sequences. This is a great achievement, one of the most important successes of twentieth-century biology. But we sometimes seem to have forgotten that the original question in genetics was not what makes a protein but rather 'what makes a dog a dog, a man a man'. It is the phenotype that stands in need of explanation. It is not just a soup of proteins.

Mapping the alternative metaphors

One of the delights of colourful metaphors is that much is left to personal interpretation. The interpretations given here are not unique or exclusive. Metaphor invention is an art not a science and, as with other art forms, the artist is not necessarily the best interpreter. I cannot guarantee that my interpretations of Richard Dawkins' original metaphors would be his, nor that the reader's interpretations of my proposed alternatives might not be better than my own. The purpose of this section is rather to illustrate the strengths and weaknesses of the statements in the context of the subject of this book. I also describe the weaknesses in the original that my alternatives seek to correct. They in turn have weaknesses of their own of course.

ORIGINAL: *now they swarm in huge colonies.* The force of this statement is that, like swarms of bees or locusts, genes are numerous individuals forming a large collective, each having its own interests and freedom to act. The colony does what it does as a result of these actions of the individuals. They 'choose' to swarm inside bodies as one of their innately 'selfish' actions.

ALTERNATIVE: *now they are trapped in huge colonies.* From the viewpoint of the organism, genes are captured entities, no longer having a life of their own independent of the organism. They are forced to co-operate with many other genes to stand any chance of survival. As Maynard Smith and Szathmáry (1999: 17) express it, 'co-ordinated replication prevents competition between genes within a compartment, and forces co-operation on them. They are all in the same boat.'

Both statements share the idea that there was a time early in the evolution of life when nucleic acid molecules (probably RNA) really were individuals in the chemical soup, when the objects of selection were molecular (Maynard Smith and Szathmáry 1999). Whether one then views the genes as 'invading' the organisms or organisms as 'capturing' the genes is largely a matter of viewpoint. Most likely, cells and genes evolved together, just as genes and proteins must have done. In both cases, neither makes sense without the other.

ORIGINAL: *safe inside gigantic lumbering robots.* This colourful metaphor resonates on many levels, but the main effect is to denigrate the 'lumbering' organism compared to its (presumably nimble) genes. The impact of the word 'robot' is powerful. Probably to Dawkins' dismay, this is one of the words that have led people to read a fully gene-determinist viewpoint into his book. A robot, after all, is entirely in the control of something or somebody else. In later books, Dawkins significantly qualifies his position in *The selfish gene*. Thus, in *The extended phenotype* he writes: 'In many cases the two ways of looking at life will, indeed, be equivalent', which implies that he acknowledges the validity of the higher-level viewpoint.

ALTERNATIVE: *locked inside highly intelligent beings*. The force of this alternative is to emphasise that any intelligence the system has is at the level of the organism, not at the level of genes. The only way to avoid this correction would be to say that the relevant 'intelligence' is already encoded in the program of the genes. I will argue in Chapter 4 that the genome is not a program and does not therefore even have this 'encoded' form of intelligence.

ORIGINAL: *sealed off from the outside world*. This refers to one of the central dogmas of modern biology. The CGAT codes for genes do not change as the organism adapts to its environment. By this criterion, the inheritance of acquired characteristics ('Lamarckism') is impossible. Fine, except that the statement suggests that organisms are defined only by their genes; whereas in truth they are also defined by the very varied ways in which genes actually operate within a living cell, and these gene expression patterns are most certainly influenced by the outside world.

ALTERNATIVE: *moulded by the outside world*. The number of possible gene expression patterns is virtually unlimited (see Chapter 2). What is more, these patterns are determined at higher levels of the organism in the context of its interaction with the environment. Moreover, the expression or repression of genes may be affected by experience in a previous generation. So, all essential characteristics of gene function except for the coding *are* 'moulded by the outside world'. Which of these statements you prefer depends therefore on whether you are focusing on gene coding or gene expression. Both are valid biological viewpoints. One (gene coding) takes a long time (generations) to change significantly; the other (gene expression) can change in hours.

There might be some reason to prefer the original statement. The environment influences gene expression levels during the lifetime of an individual organism, but the changes that result are in principle not inherited. What is passed on to the next generations is therefore 'sealed off from the outside world'. However, this is not completely certain. Some maternal effects additional to the DNA coding do seem to be transmitted. These will be discussed in Chapter 4.

ORIGINAL: *communicating with it by tortuous indirect routes*. Genes do not interact with the environment of the individual except via the proteins they code for. The proteins are then responsible for all the functional interactions.

ALTERNATIVE: *communicating with it by complex processes*. Routes of interaction could be tortuous and indirect but still be routes that are individual to each gene. I therefore prefer to emphasise that the problem is not so much the indirectness of the interaction, but rather its complexity, with the products of many genes being involved co-operatively in each functional interaction. The relationship between the two statements in this case differs from most of the other pairs in that the alternative is not so much an opposite, as an attempt to emphasise a different, and I think much more important, aspect.

ORIGINAL: *manipulating it by remote control*. This statement puts the genes in the position of 'calling the shots' by being cast as the manipulators. It also therefore contributes to the gene-determinist viewpoint.

ALTERNATIVE: *through which, blindly, as if by magic, function emerges*. The alternative de-emphasises the genes as the manipulators, emphasising rather that they are blind to the emergence of higher-level function. The phrase 'as if by magic' is optional. I add it only to highlight the sense of wonder at the beauty and complexity of what emerges (see the Silman story in Chapter 3).

ORIGINAL: *they are in you and me*.

ALTERNATIVE: *they are in you and me*. This is the only statement that I regard as unambiguously empirical, so I have left it unchanged.

ORIGINAL: *they created us, body and mind*. This statement also contributes greatly to the gene-determinist interpretation. I see two major problems with it. The first is that, even if one does think that the genes code for a program that creates us, they certainly don't do that alone. The second is that I don't believe that such a program exists (see Chapter 4).

ALTERNATIVE: *we are the system that allows their code to be read*. This alternative is absolutely central to the different viewpoint I develop in this book. There are several ways in which it can be construed.

The first, and most fundamental, is that the DNA code of a gene is nonsense (just a sequence of CGAT bases) until it is interpreted functionally, first by the cell/protein machinery that initiates and controls transcription and post-transcriptional modifications, and then by the systems-level interaction between proteins that generate higher-level function. A gene can do nothing without this interpretation by the system.

On its own, the stretch of DNA code for a gene is like a word without the semantic frame of its language. The system provides the semantic frame and gives the gene its functionality, its meaning. Equally, the system cannot exist without the genes. But there is nevertheless an asymmetry. The logic of successful systems that win in the competition for survival lies in the system, not the genes. It is systems (organisms) that live or die, not genes. What the genes do is to contain the database from which the system can be reconstructed. They are the 'eternal' replicators. They don't die, but outside an organism they also don't live.

The second way to construe the statement is that, in the interpretation of the genome by the system that translates it, there are compensatory mechanisms that enable many gene knockouts and other forms of malfunction, such as bad mutations, to be neutralised (see Chapter 8).

The third way to construe the statement is to allow that the system is not just the body itself, but also the environment. There are many kinds of adaptation of organisms to the environment, including adjustments to high altitude, to extreme cold, to poor diets and so on. Many of these adaptations involve changes in gene expression profiles.

ORIGINAL: *and their preservation is the ultimate rationale for our existence.* This is the focal polemic of the original metaphor to which the other statements lead. It relies on adopting an exclusively gene-oriented viewpoint. Once we begin to see genes more as a database for reproducing successful evolutionary experiments than as a determining program, the force that this metaphor carries is completely lost.

ALTERNATIVE: *and their preservation is totally dependent on the joy we experience in reproducing ourselves. We are the ultimate rationale for their existence*. This reversal of the original statement shows how easily one can take a completely different view. It is not for biological science to tell you which is correct. The social and ethical implications of your choice are, however, profound.

Nature, of course, couldn't care less about such questions. They are rather like the old-fashioned version of this type of conundrum: 'Which came first, the chicken or the egg?' Co-evolution is the obvious answer.

Nevertheless, it does seem to me more natural, and certainly more meaningful, to say that the rationale for existence lies at the level at which selection occurs. This is the level at which we can say why an organism survived or not.

2 § The Organ of 30 000 Pipes

There wouldn't be enough material in the whole universe for nature to have tried out all the possible interactions, even over the long period of billions of years of the evolutionary process. (this chapter)

The Chinese Emperor and the poor farmer

Over 2000 years ago, the story goes, a Chinese Emperor was saved in battle by the action of a lowly farmer. Perhaps it was the Emperor Qin-Shi-Huang, who is sometimes called the 'First Emperor' because he was the first to unite China into something resembling the huge country we know today, and he must have fought many bloody battles to do so.

Anyway, the story continues that when the battle was over and the Emperor had returned to his palace, he decided to summon the farmer for him to be rewarded. 'You saved my life. I am deeply indebted to you. I must therefore grant you any wish you have. You can have anything you want in the world.' The farmer looks around the sumptuous palace and replies, 'Sire, can you bring out your chessboard?' Of course, the Emperor has many. He commands his courtiers to find the most expensive chessboard in the palace.

When they return, instead of taking the immensely valuable chess-board as his reward, as the Emperor expected, the farmer places it firmly on the floor of the court room and then feels inside his trouser pocket to find 15 old grains of rice mixed with the inevitable dirt of his pocket. As he holds the grains in his hand, hovering over the chessboard, the Emperor looks at him bemused. 'I offer you the most jewel-encrusted, the most valuable, chessboard in the palace. And you are about to throw the filthy contents of your pocket all over it!'

'No sire, I am not going to throw these rice grains all over the board, I am going to arrange them in a particular order on your board. Watch carefully.'

The Emperor and his courtiers look even more bemused as the farmer selects a single rice grain out of the tiny pile and places it in the middle of the first square of the chessboard. The board is so large and beautiful that everyone has to look very closely to see the tiny rice grain and to distinguish it from the inlaid ivory patterns. He then takes two more grains of rice to place on the second square . . .

The Emperor thinks the farmer must be about to use the rice grains as his chess pieces. 'The man's an idiot', thinks the Emperor, as he asks the courtiers to bring out the best set of chess pieces in the palace.

The courtiers rush to bring out the pieces. They are even more sumptuously inlaid with jewels than the chessboard itself. Each piece is a unique work of art. Meanwhile the farmer continues by placing four grains of rice on the third square. He checks the remaining grains of rice in his hand; there are eight of them. He is just about to pour them, with the dust from his pocket, onto the fourth square as the courtiers place before him the glittering rows of chess men.

'These also are yours', says the Emperor. 'This is the least I can do for saving my life.'

The farmer ignores him, and the ornate chessmen. He finishes tipping the last rice grains onto the board, rubs his hands. Catching the Emperor's eye, he bows, then speaks. 'Sire', he says 'I don't need your beautiful chessboard, nor your glittering chess pieces. They will be useless to a poor peasant farmer. All I want is for you to complete the process I have begun: one grain on the first square, two on the second, four on the third, eight on the fourth. Just keep going like that until you reach the 64th square of the board. I will then take just the rice, you can keep your chessboard and jewelled pieces.'

By now the Emperor is completely disdainful. However skilled the farmer was in bodily action (for he showed great agility to save the

Emperor's life), he is a naïve country bumpkin who clearly doesn't realise the immense value of what he has just turned down. All he is worried about is the next bowl of rice for his family!

So, the Emperor makes a quick guess at what is needed and commands the courtiers to bring the largest bag of rice from the storerooms. It takes several men to bring the 100 kg bag of rice and they drop it with a large thump on the floor near the chessboard. 'Continue the process', he commands, 'as the farmer has begun—and don't bother too much about how accurately you count, for at the end I am going to give the farmer the whole bag of rice anyway.' The farmer smiles and bows.

The courtiers do as the Emperor commands, except that they do in fact count carefully. They are not palace courtiers for nothing—their first job is to be careful with the palace stores. They are not at all sure that they would be happy to see this dirty ignorant farmer go off with enough rice to feed his family for a year. So the farmer and the Emperor sit and watch as the courtiers laboriously do the counting: 16 grains on square five, 32 grains on square six, 64 grains on square seven . . . At first the counting goes quite quickly and the huge rice bag remains obstinately full.

But after the 11th square they find that they are already having to count out the rice grains in thousands. One of them then realises the wisdom of the second part of the Emperor's command. He exclaims, 'Let's use a measure that takes about a thousand grains.' So for the next 10 squares they count more approximately with the measure.

By square 16 they find that they need to count more than 30 of these measures. In fact there isn't room on the chessboard squares for what they measure out, so they start counting out piles on the palace floor, noting carefully which pile corresponds to each square of the board. At square 21, they find themselves counting 1000 measures. Worse still, at square 22 they run out of rice. The whole 100 kg bag is suddenly used up as they reach the more than 3 million grains required for the next square. The piles of rice now extend well beyond the chessboard. They look appealingly at the Emperor. 'What do you want us to do, Sire, do we get more bags?'

The Emperor looks a little puzzled. He is still thinking in linear terms. It can only take a few more bags, he says to himself. 'OK, let's get another 10 bags out of the store, then continue.' Anyway, the Emperor has already begun to lose interest. One of his most beautiful concubines has appeared at the back of the audience chamber.

But, for one of the courtiers the penny at last begins to drop. He whispers to his colleagues: 'It won't work. Just think about it: if you add together all that we have placed on the first 22 squares, which is already more than a whole bag, we will have to put that quantity again on square 23,[1] so that's more than one bag on just a single square! And then the next square after that will require more than two bags. We will need hundreds of bags to finish the job!'

The Emperor overhears this. Despite the concubine's alluring presence, he is also beginning to get worried, but the courtier's reference to hundreds of bags starts to reassure him. 'No problem', he says, 'there was a bumper harvest. On the last stock-taking, we had thousands of bags in the palace store. Go! Bring as many bags as you need.'

They reach square 32: just halfway through the chessboard. More than 300 of the palace storeroom rice bags have been used up. It will take 300 more to go to the next square, and then 600 bags for the one after that. They are no longer piling up bags on the palace floor; they are down in the storeroom labelling bags with chessboard numbers.

The mathematically literate courtier does a quick calculation. He shows on a piece of paper that by square 36, all the rice in the palace will have been exhausted; by square 50 all the rice in China will not be sufficient; by square 64 they would be filling the whole surface of the world more than knee-deep in rice.

He goes back up to the palace court-room and whispers his calculations to the Emperor, showing him that to finish the job would require a fortune billions of times greater than the imperial possessions. The Emperor goes white with shock. He dismisses the concubine and looks as though the world has suddenly fallen in. To satisfy the farmer's humble request he will be ruined.

The farmer realises that, at last, the Emperor understands.

---

[1] He's understood that the effect of doubling is to add together everything that has already been put on previous squares, plus one grain.

'Sire,' he says, 'you are a great and powerful man. You proudly promised to give me anything in the world. But with the simple means of a chessboard and 15 grains of rice I have been able to teach you a lesson. Even you do not understand the immensity of the world. You cannot promise what you do not have. But don't worry. I will just take all but the last bag of rice in your storeroom. You can keep the last one. That will feed *your* family for a year! You really don't need the rest.'

The Emperor never forgot the lesson. In his rage he made 700 000 convicts work for the next 38 years to construct the terracotta warriors that were to defend his tomb. After 30 years of digging, the modern Chinese archaeologists have still only explored a part of the approaches to the tomb. The tomb itself has yet to be opened. But already they have over 7000 warriors on display in a spectacular exhibition near Xi'an, the contemporary name for the Emperor's ancient Chinese capital Chang-an.

The genome and combinatorial explosion

I hope this story did not shock you as much as it did the Emperor. For you will find much greater surprises in store.

Start with one, double it, then double the result, and so on 64 times—and you end up with a very large number indeed (100 billion trillions). But suppose we expand the size of the chessboard. Instead of 64 squares, let us say we had 30 000. That is the number of genes in the human genome, according to one current estimate. And then let us change the way we increase one number to produce the next. In the story, the rule was: count one pile of rice and then double that number to get the size of the next pile. In this case, let us instead calculate how many interactions are possible between one set of elements, and let that number be used to give us the size of the next set.

In other words, mathematical series show differing degrees of non-linearity. Some converge to a particular number, some can blow up quite gently, yet others display a form of mathematical explosion. It is the latter type that we are dealing with here: one of the most non-linear of mathematical functions. Combinations of elements grow very rapidly indeed as the number of elements increases. We call this 'combinatorial explosion'.

In Chapter 1 we learnt that genes, or rather their products, the proteins, interact in large groups to generate biological functions. There are no biological functions that depend on the coding provided by a single gene. How many, then, do we need? We are far from knowing the answer to that question. But we do know that nature is modular, so that it makes sense to refer to a particular group of genes or proteins as acting in at least semi-isolation from the rest, to generate a function. I will discuss this feature in later chapters. How large might these function modules be? We can make some lower-level guesses.

First, let's ask the following somewhat absurd question: if just two genes were required to co-operate to generate a biological function, how many possible functions would there be for a genome of 30 000? The answer is $(30\,000 \times 29\,999)/2$ which is $449\,985\,000$.[2] So, even with this minimal number of genes per function, there could be nearly 500 million different biological functions.

Now let's be a little more realistic. In Chapter 5 I will show that we can model one of nature's most important oscillators, the pacemaker rhythm of the heart, with less than 100 biological components, that is, functioning proteins. Successful simulation of some biochemical metabolic networks, such as those found in bacteria, can also be done with a similar number of components.

So let us suppose 100 genes were required for each biological function. What then would be the possible number of functions in a genome of 30 000? The answer is truly gigantic: approximately 10^{289}! For comparison, the game with the chessboard in the story generated a number of 'only' 10^{19}.

---

[2] This element in the series we are investigating is fairly simple to understand. Each gene, that is, 30 000 could, on this scenario, interact with each of the remaining genes, that is, 29 999. Half the number we calculate in this way would be redundant since we are assuming that gene x interacting with gene y is the same as gene y interacting with gene x. So, we have $n(n-1)/2$ as the number of potential interactions.

And what if we now remove the 100 gene restriction and allow *any* combination of gene interactions to generate a function. Then we get $2 \times 10^{72\,403}$ (Feytmans *et al.* 2005). That is a number with over 70 000 digits. It would require about 30 of the pages of this book just to write the number down.

These numbers are so large that there wouldn't be enough material in the whole universe for nature to have tried out all the possible interactions, even over the long period of billions of years of the evolutionary process. Given their enormity, and their physical impossibility, they have no realistic physical meaning. They are not just upper limits, they are way beyond any possible upper limits. This hypothetical procedure is a complete non-starter.

And here's the rub. Much contemporary popular writing on genetics assumes that it should be possible to reconstruct living systems from the bottom up, starting with the raw DNA code. And that is precisely the sort of procedure we have just seen to be so entirely impracticable. Clearly, we need first to narrow the options. And there is only one way to do that; we must observe how nature itself has narrowed the options.

It is clear that nature has had vastly more theoretical possibilities than are actually manifested in existing species. And, after all, that is what we should expect. It is hardly sensible to assume that any given protein should be able to interact with all other proteins (and with all the metabolites and signalling molecules in the cell as well). The chemistry of protein–protein interaction will impose limits here. Different proteins play different types of role. Some are hubs; they occupy central positions within networks. Others are relatively unreactive, occupying peripheral positions.

Location is also a major limiting factor. Within cells and organs, there are many compartments; each is relatively isolated from every other. So we have to consider what proteins are found in what compartments. Then, where each protein occurs within the compartment is also important. Some sit within membranes or organelles and can only react with molecules that can access the spaces around them.

On that basis, we can get a much clearer picture. Some studies have already been done. One focuses on the bacterium *E. coli*. It estimates the total number of possible networks within this organism's metabolism at around 4.4×10^{21}. The number of circuits that are actually used appears to be tiny in comparison—half a million or so. In this case, then, it looks as though there are 10^{16} times as many possible circuits as there are circuits actually in use (Stelling *et al.* 2002).

So only a small proportion of the ways in which the genes could theoretically 'be expressed' are in practice ever going to happen. In fact it is an even smaller proportion than we have so far established, because the number of possibilities is larger.

Our calculations so far are based on each gene being a single fixed entity. In reality, many different equivalent forms (called isoforms) of each gene are available. This will further increase the potential combinations. Then there are the splice variants: most genes have more than one exon so that there are at least three possible splice variants. Then there are the processes that happen after the information is read off the DNA: post-transcriptional regulation, in the vernacular.

We have already met the *Dscam* gene in the fruit fly *Drosophila melanogaster*. This has been shown to have as many as 115 exons, with 38 000 putative protein products, some of which change their expression levels between birth and adulthood (Celotto and Graveley 2001). It may also be responsible for the extensive immunity system of insects (Watson *et al.* 2005).

Another important conclusion is that the total number of possible functions increases much more rapidly than the total number of genes. Compare, for example, 500 genes with 5000. At 100 genes per function, we have more than 1×10^{100} possible functions for 500 genes, and over 1×10^{200} for 5000 genes. Thus increasing the gene number by one order of magnitude leads to multiplying the available combinations by a further factor of 10^{100}. So, one order of magnitude increase in gene numbers generates 100 orders of magnitude in potential functions. The dependence of potential functions on the number of genes is therefore highly non-linear.

The significance of this result becomes even clearer when we compare genomes of similar sizes or of great similarity in sequence. Consider, for example, the effect of adding one gene to a genome of 30 000 genes. The number of possible new potential functions is about 10^{287}. Conversely, if a function requires just one more gene, then the number of newly created possible combinations would be about 10^{292} (Feytmans *et al.* 2005). These numbers are so unimaginably large that we can invert the usual question that is asked when genomes of great similarity are compared. The question should not be 'How can this very small difference possibly code for all the functional differences between the species, or for the increased complexity?', but rather 'How many of these immensely large potential differences does nature actually use and how are they chosen?'

An organ of 30 000 pipes

In Chapter 1, I proposed that the genome is not at all like a program that 'determines' life. It is more like a CD, a digital database that stores information to enable something to be reproduced.

In this chapter we have wondered what to make of the fact the human genome includes around 30 000 genes. Should we be surprised that there are so few? Should we not rather be amazed at the immense range of functional possibilities that such a genome can support?

A musical analogy may be helpful here. The genome is like an immense organ with 30 000 pipes. Pipe organs were developed to enable the impressive music to be played for which organs are so famous. The larger the organ, the more pipes it has, and the wider the range of pitch, tonality, and other musical effects it can be made to produce. The music is an integrated activity of the organ. It is not just a series of notes. But the music is not itself created by the organ. The organ is not a program that writes, for example, the Bach fugues. Bach did that. And it requires an accomplished organist to make the organ perform.

By happy coincidence, 30 000 is about the number of pipes in the world's largest organs.[3] Even much smaller organs, as in most

---

[3] The record is held jointly by two organs in the USA. The Wanamaker organ in the Lord and Taylor department store in Philadelphia has 28 482 pipes, and 396 registers. It was first built by Wanamaker's organ shop between 1914

cathedrals, enable a vast range of music to be played. One of 30 000 pipes can, surely, permit the whole music of life to be played.

If there is an organ, and some music, who is the player, and who was the composer? And is there a conductor?

These are questions for the next chapters.

and 1917. The Convention Hall Organ in Atlantic City has 33 114 pipes with 337 registers. These are both around the size of the human genome, given a likely error of ±10% in estimating the number of genes. The largest pipe organ in the UK is that in the Royal Albert Hall, with 9999 pipes, which is similar to the number of pipes in the Sydney Opera House. Notre Dame Cathedral in Paris has about 8000 pipes.

3 § The Score: is it Written Down?

I believe very strongly that the fundamental unit, the correct level of abstraction, is the cell and not the genome.

Sydney Brenner, lecture at Columbia University, 2003

Is the genome the 'book of life'?

We have seen that the human genome is a vast database of information, 3 billion base pairs, which make up 20–30 000 genes. Each of these is used to encode the amino acid sequence of a particular protein, or set of proteins. The complete sequence and structure of the proteins is sometimes referred to as the proteome. So how is the information in the genome used to create the proteome? This is quite a challenging question.

First, we need to identify all the genes, which we are getting closer to doing, at least in the sense of identifying which parts of the DNA code correspond to genes. Still, that doesn't mean that we know what each gene does, what is its functionality.

Second, we now know that there are many more proteins than genes. So what determines which protein is made, when? Some sort of feedback control, obviously—but, what sort? There is a complex interaction between genes and their environment—both the cellular environment and also the wider environment of the organisms in which they exist. The organisms in turn interact with their environment, and this also will have an impact on gene expression. Clearly, the simplistic view that genes 'dictate' the organism and its functions is just silly. It avoids the real challenge we face here, which is to understand the control processes that determine which proteins are produced ('expressed') and to what extent.

And *third*, amino acid sequences are revealing, but do not explain all we need to know. We can know a lot about a protein in terms of its amino acids, and still struggle to work out what three-dimensional structure it will form, and what chemical function it will perform.

At first, investigators of the human genome thought there were many more genes than we have in fact found. Early estimates were of the order of 150 000, whereas we are now looking at between 25 000 and 30 000. This discovery adds to our difficulty. It means that each gene must very commonly be involved in many different biological functions. Indeed, such multiple functionality is probably universal.

These are formidable challenges. Still, they pale into insignificance when we consider the complexity of the next stage. This is when we have to try to understand the interactions of the proteins. How do the tens of thousands of different proteins interact? How do they thus generate or at least contribute to what happens ('functionality') at all levels of the organism, through the cells to the organs and the wider biological systems? This is the task of quantitative analysis of physiological function, which in its entirety is sometimes now called the physiome.

Clearly the build-up from the information in the genome to the living systems it helps sustain is immensely complex. There will be many stages along the way of unravelling it. Mathematical modelling of physiological systems combined with the new field of bioinformatics will play an increasingly important role here, as I will show in later chapters.

From a systems biology viewpoint, the genome is not understandable as 'the book of life' until it is 'read' through its 'translation' into physiological function. My contention is that this functionality does not reside at the level of genes. It can't because, strictly speaking, the genes are 'blind' to what they do, as indeed are proteins and higher structures such as cells, tissues, and organs.

To these I want now to add two more important points. Proteins are not the only molecules in biological systems that determine function. Function is also dependent on the properties of water, lipids, and many other molecules that are not coded for by genes.

Moreover, a lot of what their products, the proteins, do is not dependent on instructions from the genes. It is dependent on the poorly understood chemistry of self-assembling complex systems. It is as though the genes specify the components of a computer, but not how they should be put together. They just do this by doing what is chemically natural to them. Some people are predicting that this is also the way in which we will build computers in the future, particularly what are called molecular computers.

So, if the genome is 'the book of life' it is a book with enormous gaps, which nature takes for granted since it never had to work out how to code for such natural phenomena. There are no genes for the properties of water, or for the fatty lipids that form cell membranes. Worse still, as we shall see in later chapters, there are no genes specifically coding for interactions. All of this 'missing information' is implicit in the properties of the environment in which genes operate. Water, for instance, is a wondrous substance that behaves in many remarkable, intricate ways—and the way living organisms develop and function is conditioned by, and reflects all the characteristics of this key material.

The same goes for lipids. These are the fats of the cellular world—they don't dissolve in water. One type, the phospholipids, is a major component of plant and animal membranes. If we ask how an organism manages to grow, one of the answers is: because lipids behave as they do.

Moreover, this environment crucially determines which genes are expressed and to what degree. The passage of information is not simply one-way, from genes to function. There is two-way interaction.

The French bistro omelette

There was once a little family bistro outside Paris. It acquired a reputation for the great lightness, flavour, and delicacy of its omelettes. A group of connoisseurs were writing up a compendium of French cuisine. They decided it was essential to include a recipe for these famous omelettes.

The mother of the house kindly obliged. She gave them a detailed recipe of ingredients and the order and manner in which they were blended as the omelette was prepared. There was only one snag. When the Parisian chefs tried it, they got a different result. They found delicious flavours but their omelettes were totally lacking the lightness that the little family bistro achieved.

Frustrated, they experimented. They interpreted the recipe in various ways. They tried different forms of omelette pan. Eventually, they decided that there must be a trick. The mother had surely not revealed all her secrets!

So, off they go to the bistro. There, they find that the mother has died. Her daughter now cooks the omelettes.

However, her omelettes are just as good as her mother's. So they show her the mother's recipe and ask whether it is correct. She reads it carefully. Yes, she says, it is a marvel of accuracy, down to the last milligram of ingredient. This is exactly what she herself does. 'There's nothing missing?', they ask. 'Of course not', she replies, 'Mother has written it all down.' In fact, she says, her mother wrote it down for her exactly the same way. She is following precisely the same recipe as they were trying to follow.

They naturally ask her whether they can watch as the omelette is prepared. Of course, she says. There is nothing to hide. So they watch carefully to try to detect the slightest difference between the written recipe and what she does. And they are amazed at what they see. At the very beginning of her preparation, she separates the whites and the yolks. She blends the flavourings into the yolks and folds the beaten whites in only at the end, just before cooking.

They upbraid her. She and her mother, they say, have both provided inaccurate information. How could anyone fail to reveal this crucial fact in such a recipe? The daughter is offended. She takes them to task for their 'stupidity' and 'arrogance'. Indignantly, she surveys the assembled dignitaries and asks them, in all innocence, 'How else do you think anyone prepares an omelette!?'

Similarly, nature has not coded for what is chemically natural to proteins. It does not need to. And this information has at least as much of a claim to be 'the book of life' as does the genome. Moreover, it is going to be much more difficult to work out this side of the story than to sequence the genome.

So, in brief, my view is that the genome is most like a tedious machine code for the construction of the key players in the game of life: the proteins.

It is incomplete in a major respect: how these proteins behave chemically in the cells of the body, how they fold, combine, and interact. It is also completely lacking in functionality. It does not even tell us whether a particular gene plays a role in one, two, three, a dozen, or a hundred functions. And, finally, it relies time and again on Mother Nature's ability to know 'how to make omelettes'.

The ambiguity of language

One defence of the 'book of life' view against my criticisms is to say that all books are like this to some degree. All languages function in a context of implicit knowledge that the language itself does not need to spell out. Of course this is true. One way to see this is to note that languages differ in what they take for granted.

Some languages do not normally use plurals, for example, though equally obviously the users of these languages—which include Chinese, Japanese, Korean, Polynesian languages, Maori—have the concept of plural and know when there is more than one of the thing being referred to.

There are innumerable examples of words, even when very similar in different languages, that function differently simply because of the cultural context in which they are used.

To call a French woman, in French, 'séduisante' is to compliment her highly. To call an English woman 'seductive' could be highly dangerous. But to avoid the cultural difference by translating 'séduisante' as 'beautiful', 'appealing', etc. is to remove precisely the sexual frisson that the French language intends. It also begs the question of what the cultural context of the word 'sex' is in the two languages.

If such differences can be so large between two European

languages, they can become great gulfs of misunderstanding between widely separated languages. Early Japanese visitors to Europe around 1900 had to refer to people 'licking' *(nameru)* each other, or to 'lips connecting' *(seppun)* to describe what they saw since they were amazed to see people kissing each other as a form of public greeting.[1]

So, yes, clearly this kind of 'implicit' knowledge is an irreducible feature of all languages. None are, or can be, neutral with respect to culture since they are themselves the creatures of that culture. But the relationship between a language and its culture is not like the relationship between the genome and nature. One way of seeing the difference is to ask what the language is *trying* to do.

Human languages aim to describe the world as it is (or rather, as it is seen by the speakers). It is an aim of each language to try to avoid ambiguity. There is even a discipline, called philosophy, that— working as it must within the constraints of a particular language— tries to step back, as it were, and see beyond the limits of the culture, to question those limits, the bounds of sense (to echo Kant) and of meaning.

By contrast, the genomic language does not have anything like this as an aim. Of course, strictly speaking, it must have nothing as an aim—genes, and evolutionary processes are blind (so also are cells and organs, but we will come on to them in later chapters)—but if we take this line then, with even more certainty, the genome is not a book. In order to pursue the book analogy we have at least to ask what it could be said to be a book *about*. So is it about 'life'?

Well, I would say no, not really. Crucially, it does not describe functionality. The code for gene XYZ does not spell out that XYZ is the sequence for a protein that enables synapses in brains to function, testicles to produce sperm, pancreatic cells to secrete insulin etc., etc. This is a bit like imagining a book that does not spell out that X is the

---

[1] Modern Japanese has a verb, *kisu-suru*, which borrows from the English 'kiss'. The misunderstandings occurred both ways. Some Westerners even thought that the Japanese never kissed each other. There are related problems with exploring the conceptual frames for the words for 'love' and 'sex' in the two languages. Prior to twentieth-century cultural globalisation, an East Asian person would not dream of using a sentence like 'I love you' (Downer 2003).

king, Y the archbishop, Z the villain, etc. But, worse than that, from reading the 'book' we don't even know what the relationship between a king and an archbishop might be. These crucial interactions are outside the scope of what the language of the genes specifies.

The Silmans return

So, let's do a thought experiment.

Silmans, you will recall, are an intelligent species that evolved in a world in which silicon has the functions of carbon. They advance to the point of rapid space travel and find the planet earth. But, they have a serious problem. They can't live on earth. The earth's environment is terribly hostile to Silmans as a form of life. Moreover, they can't bring humans, or any other earth creatures into their spaceships, for what they require to live is equally hostile to earth forms of life.

But they know from their own evolution that there must be something equivalent to Silman code, there must be earth-life code. They have also worked out that, in their case, this is imprinted in chemical sequences that are inert. Their code molecules do not breathe, or need siligen (the equivalent for them to oxygen), etc., etc.

They therefore reason that they might be able to send robots down to the planet's surface that will extract earth-life code. OK, let's call it DNA. To their great satisfaction, they find that their surmise was right. They find that this DNA is also chemically inert. It can be taken into their spaceships and it can be subjected to analysis so that the sequences can be read. They get right down to it. Soon they have read the complete DNA of a human.

Let's give them, too, the intuition that this 'inert' molecular sequence codes for another sequence, that of the proteins, and that these are highly reactive. So, having worked out what the code means (which DNA sequence corresponds to which amino acid) they set about determining all the proteins—all 100 000 or more of them— that go to make a human.

But, then they are stuck. After all, theirs is a silicon world. They do not have water. They do not have lipids. But, they guess from analogy with their own world that there must be such things. They try to see in the DNA code any clues to what these substances might be. To their intense frustration they find there is nothing. All the DNA code does is to specify one type of molecule, proteins.[2] No other information is there. What a bore!

They decide that earth-life must be a very strange thing: just tens of thousands of proteins thrown together. Perhaps humans are a kind of soup! Perhaps earth-life is extremely primitive. Yawning and stretching, they prepare to time-travel on to the next inhabited planet.

Then, one of them has an idea. 'Hey, wait a minute', he says. 'You could say that we are "just" a bunch of silicon sequences. But we know we are not: that we think, reproduce, etc., etc. Evidently, there is more to life than molecular sequences.'

'What we should do is to send down another robot with a carefully isolated capsule into which it will put all this strange stuff (i.e. water, air, lipids, etc.) that we know is down there. Let's bring it all up in the capsule and see what happens when we let the DNA do its thing in this environment.' So, they do. Then, they watch what happens. With incredulity, they observe how cells are formed, then divide into the early embryo, which then goes through all its wonderful transformations until nine months later a human emerges.

Well, all right—we are simplifying grossly here. They are just cloning—and there's no mother, no womb. This is a test-tube baby in the fullest sense of the term. Even so, we have to assume that they picked up at least one cell to provide the environment in which the nuclear DNA can work. Still, these are details. The point of the story remains. How, we can ask ourselves, do the Silmans react to this remarkable experimental result? Presumably, just like the Parisian chefs, they feel cheated. The 'recipe' doesn't specify all this. It just happens!

---

[2] This simplification is not entirely true. Some DNA codes for RNA molecules that do not code for proteins. These RNA molecules, together with around 100 different proteins, form the cellular machinery, called ribosomes, that strings together the amino acids to form proteins. I have ignored this complication since all DNA is first coded into RNA, some of which is then used by the ribosomes as the template for making proteins.

Some omelette!

The point is this. Much more than the genome is involved in the development of an organism. If there is a score for the music of life, it is not the genome, or at least not that alone. DNA never acts outside the context of a cell. And we each inherit much more than our DNA. We inherit the egg cell from our mother with all its machinery, including mitochondria, ribosomes, and other cytoplasmic components, such as the proteins that enter the nucleus to initiate DNA transcription. These proteins are, initially at least, those encoded by the mother's genes. As Brenner said, 'the correct level of abstraction is the cell and not the genome'.

Also (a mere bagatelle, I allow), we inherit the world. The peculiar chemistry of water, lipids, and many other molecules whose form and properties are not coded for by DNA—all that is given. And yet the central biological dogma of our times is that inheritance is solely through DNA! How has this remarkable notion taken such a strong hold? What are the implications of breaking free from it? These are questions we must address.

4 § The Conductor: Downward Causation

> Organisms are not simply manufactured according to a set of
> instructions. There is no easy way to separate instructions
> from the process of carrying them out, to distinguish plan
> from execution.
>
> Coen 1999

How is the genome played?

Who plays the organ of 30 000 pipes? Is there an organist? What sort
of organist could there be?

The organist works from a very different perspective to that of any
of the individual pipes of his organ. Although in real organs the pipes
themselves are often physically above the player, metaphorically
speaking he surveys the keyboards and pedals from above, seeing patterns and musical forms that he constrains the instrument to follow.
To play a piece on the organ is to call on many pipes simultaneously.
In a complex piece, most of the pipes may at some point come into
play. The organist must cause this massive apparatus to express its
potential in an ordered way. These ordered patterns form a very tiny
fraction of all the possible patterns that could be played, the great
majority of which would be cacophonic.

There are clear similarities to the way in which the genome is
played. To form a high-level physiological function, such as the
heartbeat or neural function, large numbers of genes are expressed
simultaneously. Very probably, as much as a third of the genome,
10 000 genes, may be expressed in an organ like the brain. Similar

numbers may be involved in other organs and systems, like the heart and the liver. With 'only' 20–30 000 genes, that may appear to be problematic.

However, we don't run out of genes because many are re-used to be expressed in more than one organ and system. Many genes are expressed in *all* the organs and systems of the body. It is the pattern of expression, not the individual genes, that defines which kind of organ or system is formed.

All the genes in the genome are present in the nucleus of every cell in the body, but many are switched off, involving different groups in different organs, and even those that are switched on show large variations in the amounts of protein formed according to which organ or type of cell they find themselves in. Over the course of a lifetime, these organ and system gene expression patterns also change. The embryo, the young baby, the growing child, the athletic young-ster, the busy parent, and the aged grandparent all show significantly different expression patterns.

The music of life is a symphony. It has many different movements. Some melodies find echoes in more than one, but the movements are nonetheless distinct.

Is the genome a program?

Remember the simple diagram in Chapter 1 (Fig. 1) that shows the bottom-up, reductionist view of life? On this view, the genome dic-tates to all the other levels. That works in certain limited contexts, of course. But it does make it difficult to understand all the complexity we are now seeing.

In a purely bottom-up perspective, the genome is viewed as dictat-ing to all the other levels. Some people do in fact refer to the genome as the recipe for life, as though it were a set of instructions that causes everything to happen in the correct sequence. On this view, there must be a representation of the possible forms of the organism in its genome. This representation may be difficult to read, but it is thought to be there, encoded in a program that slowly unfolds.

In a way, this is not so much an interpretation of biological data as the reflection of pre-existing assumptions about how things are meant to be generally. These assumptions have a history, in biological science as elsewhere. Thus, earlier biologists imagined that the germ cells must contain miniature versions of the adult organism. Whether the program is an actual map (an organism in miniature) or a coded version of this map that needs to be interpreted doesn't change the basic idea. What comes into being, on this view, must all have been tightly defined in some minimal form beforehand.

Of course, people who think this way also allow that the environment exerts an influence. But this influence is seen as essentially the fine-tuning of a process that is primarily inherent in the genetic make-up of the individual organism. This way of thinking leads to some strange results. People try to assign different fractions of influence to nature and nurture. If genes could be held responsible for, say, 60% or even 90% of what we are and do, then, it is thought, we could say we have an organism that is by and large genetically determined. Oh, yes?

Of course, it is true to say that the behaviour of a spider or an ant, for example, exhibits much less variability than that of a monkey or a man. So we could say it is much more genetically fixed. But does that mean that there is some linear scale of outcomes, which we can arrange in a 0% to 100% range? Surely not! Without genes we would be nothing. But it is equally true to say that with only genes we would also be nothing.[1]

Moreover, the interactions of gene products, proteins, as they form biochemical networks, are highly non-linear. So it is hard to see how we could possibly interpret any overall numbers on a linear scale. We are not dealing with interactions in which $2 + 2$ must always make 4. Here, $2 + 2$ may not only make 5, it may just as well make 105!

Nor can we avoid this difficulty by using multidimensional scales,

---

[1] There is a limited sense in which such percentages can be meaningful, which is that we can sensibly ask how much of the *variation* in some characteristic or behaviour is attributable to genetic variation in the population. This is what most of these kinds of study try to measure. My criticism is therefore aimed at those who slide from measures of variation to conclusions about degrees of genetic determinism.

one for height, another for weight, another for IQ, skin colour, hairiness, sexuality, etc. Those who think like this resemble the Emperor in the story. They think in linear terms. This is quite evidently a mistake. But it seems unavoidable once a person buys in to the basic explanatory framework of reductionism. There is something deeply wrong here. It is not just a problem of sometimes getting the numbers wrong. We need to think very carefully about this whole idea that the genome can be seen as in some way equivalent to a computer program.

Would it make sense for the genome to play a role in the living organism like that of a program in a computer? Is that what nature would have to do, to get the results that we see? Would it not be easier for nature to work as a musical composer does: simply note down enough information to enable a competent musician to recreate and interpret the piece. And take the rest for granted. At a stroke, that would greatly reduce the volume of data that must be transmitted from one generation to the next. There would be no need for a complete map of the organism, just as a musical score is not a complete miniature map of the music itself. All that is needed is a database that is sufficient, in the right environment, to trigger what is required.

Of course, that approach would only work if the player of the genome were sufficiently competent. So how does that player, the cell genomic reading facility, measure up? More than adequately! After all, the genes by themselves are dead. It is only in a fertilised egg cell, with all the proteins, lipids, and other cellular machinery inherited from the mother, that the process of reading the genome to initiate development can get going. At least 100 different proteins are involved in this machinery, without which the genome would express nothing. So, even at the very beginning of a new organism's life more is happening than is dreamt of in the reductionist's bottom-up model. The higher levels (as per Fig. 1) trigger and influence actions at the lower levels. We can call this 'downward causation'. This is how the protein and cell machinery works to stimulate and control transcription and all the post-transcriptional modifications. This is what 'plays' the genes.

Any well-regulated biological system must involve feedback controls. Clearly, then, the expression of a gene (in the rather misleading jargon) will involve levels of activity that are determined by the system as a whole. This is so obvious that it is truly extraordinary that there should be such great and repeated need to point it out.

Even at the very beginning of the life of a new organism, there is 'downward causation' involving the higher levels of Fig. 1 triggering and influencing actions at the lower levels. In fact, contrary to the view that we start life as a bunch of genes inherited from, and separated from, our parents, we start life subject to influences exerted on our genes by the environment, including the complete system of the mother.

Control of gene expression

Let's look at an example of downward causation. Nerves and muscles in the body work by transmitting electrical signals. Like all cells, a nerve or muscle cell is enclosed in a membrane. At any one time, there is a given electrical potential across that membrane. For the nerve or muscle to operate, it is necessary for that potential to change. To trigger such change, an electrical charge must be transmitted across the membrane. That charge takes a physical form—ions. Ions are molecules, or bits of molecules, which carry an electrical current. So, a stream of ions transmits an electrical charge. Such a stream has to pass down some sort of channel. This takes the form of a protein molecule. For each such protein molecule, there is at least one gene that codes for it.

One of the ions involved in this sort of process is the sodium ion, the positive charge formed from common salt, sodium chloride. The corresponding proteins are therefore called 'sodium channels'. How fast the potential across a membrane changes depends on how much of the protein there is around. If many sodium channels are expressed, the electric potential can change very quickly. This large current flow then causes a wave of excitation to spread rapidly along the nerve or muscle fibre.

Now, one of the functions of a nerve is to transmit signals very rapidly. So, one might expect that the transcription machinery would function so as to express as many protein channels as possible. But it does not.

Around 50 years ago, Alan Hodgkin worked on the equations for nerve propagation. He found that, yes, if the density of sodium channels in a nerve increases, the electrical impulse is conducted faster—but only up to a certain point. Once this point is reached, adding more sodium channel proteins actually causes the transmission to slow down. So the best way to keep the nerve functioning smoothly at its best rate is to maintain a steady rate of sodium channel expression up to this optimum level but no further. And this is in fact what normally happens.

When the system starts to get clogged up with too many sodium channels, less of them are expressed. So something is happening at the higher, system level, which causes behaviour to change at the lower, gene level. This is what neuroscientists call electro-transcription coupling. It is a form of causation that operates from the top down, not from the bottom up. It applies to all the genes that are expressed in the nervous system.

Change the frequency with which a nerve is excited or synapses are used, and the gene expression levels will change. The nerve cells feed back to their own nuclei, and this feedback controls the behaviour of the genes in the nuclei (Deisseroth *et al.* 2003). This is an ongoing, continuous process. The organ player never stops playing! The music of life lasts as long as life itself.

The same sort of thing happens in other organs, like the heart. There, the process is referred to as remodelling. The heart of an athlete, or at the other extreme, that of someone who is subject to heart failure, will show different gene expression patterns from those of an average healthy person. What is striking is that this remodelling involves very large numbers of genes. No single gene, nor even any small number of genes, can be used to fully define an athletic heart. A small number can be treated as markers in the sense that measuring them alone may be sufficient to tell us which kind of heart we are dealing with, but that is a little like saying that we can identify a whole symphony by its opening bars of music. The first four notes tell us when Beethoven's fifth symphony is to be played, but the symphony itself is much more than the first four notes!

Downward causation takes many forms

A mother transmits to the embryo adverse or favourable influences on its gene expression levels. These can determine the eventual adult pattern of health and disease many years later. These influences, called 'maternal effects', can even extend over several generations. The genome alone, therefore, does not carry all the information that a mother transmits to her offspring. That is to say, some acquired characteristics can be inherited and passed on for a generation or two. Inheritance of this kind forms no part of neo-Darwinian theory. On the contrary, it is close to the great taboo that is called 'Lamarckism'.[2] Strictly speaking, according to the standard biological dogmas such things cannot be.

A lot of effort is now being devoted to exploring such effects (Jablonka and Lamb 2005; Colvis *et al.* 2005; McMillen and Robinson 2005). We are at the beginning of what may be a long and exciting process of discovery. Mediated through the mother in a variety of ways and through the father in the germ-line, these effects are comparable to the way in which characteristics are induced in an individual during development (Chapter 7), but with the significant difference of being transmitted across the generations. Natural selection can act on them since patterns of gene marking induced by the parent and/or environment could be selected for. But it is far too early to tell whether a new form of 'Lamarckism' is waiting in the wings to creep back into mainstream biological thinking.

Some of these epigenetic effects can be transmitted through at least four generations in the case of male fertility in rats (Anway *et al.* 2005). Yet we have sought dogmatically to limit nature to just one mechanism for transmission of inherited characteristics. Clearly, we

[2] For reasons that I will explain in Chapter 7, I use 'Lamarckism' in quotes. It is the word that has come to be used for this phenomenon, but it is not historically correct.

should be more cautious. DNA does not come to us in a 'pure', unalloyed form. It must necessarily be inherited together with a complete egg cell. So any environmental or maternal effect that can influence the egg cell and/or early embryo might in principle imprint itself on the genome, or even be handed down in parallel with the genome.

Thus the gene–protein network that is involved in early development is 'controlled' by maternal proteins (coded by several maternal genes) that have entered the egg (Coen 1999; Dover 2000).

We are already familiar with a common form of such parallel transmission. The energy factories in cells are called mitochondria, and they are thought to be derived from bacteria-like organisms that were 'captured' by cells (or we could say that they 'invaded' cells). Cells and their invading mitochondria then entered into a mutually beneficial co-operation. Reflecting this origin from formerly independent forms of life, the mitochondria have their own DNA. There is therefore inheritance of mitochondrial DNA, which is not part of the nuclear genome. It is quite plausible that other parts of the egg cell protein mechanisms should do the same.

Such effects are already established in lower animals. Changes in gene expression can be transmitted through the egg in *Daphnia*; and in ciliate protozoa, patterns of cilia can be transmitted independently of DNA (Maynard Smith 1998). We will return to this question in Chapter 7.

Other forms of downward causation

Downward causation is not limited to effects on gene expression and transmission of influences through the egg cell. Any of the higher levels of the diagram shown in Fig. 1 can influence activity at any of the lower levels. The cells and organs of the body produce many different messengers that transmit these influences. These messengers are small molecules. For long-distance communication the messengers are sent through the bloodstream and are called hormones. For short distance communication between adjacent cells they are called transmitters. In both cases they act on other cells by attaching themselves to proteins on the cell surface called receptors.

In this way, hormones circulating in the blood system can influence events inside the cells through actions on cell surface receptors, which then cause chemical reactions inside the cell. The organs that produce these hormones are called endocrine glands. They can target specific cells in other organs because the relationship between the hormone and the receptor is like that between a lock and a key. Specific keys (hormones) are required to enter particular locks (receptors). Only the correct key can enter the receptors. Once that happens, a whole chain of events is triggered inside the cell that alters its function.

These are the ways in which, for example, an egg is triggered to leave the ovary, hearts are made to beat faster during exercise, breasts are triggered to produce milk, sugar is stored in cells . . . the list of endocrine functions is a long one. Biologists refer to the complete set as the endocrine system.

Transmitters are used by the nervous system to influence all the lower levels. Because the distance between a nerve ending and the organ it attaches to is very short, communication using transmitters is much faster than endocrine control. This is the way in which, for example, the nervous system rapidly causes muscles to contract, or the heart rate to accelerate.

We can see that Fig. 1 is therefore woefully incomplete. Figure 2 is an attempt to remedy this defect. Large downward-pointing arrows represent some of the forms of downward causation. There is nothing mysterious about downward causation. It follows the usual rules: one event leads to another, that leads to another, etc. Upward (reductionist) causation is no more rigorous, nor does it have any greater a claim to be scientific. We can quantify both types of causation equally. Neither is easier to encapsulate in mathematical equations.

In fact, there is nothing particularly new about downward causation. Complicated systems generally tend to regulate themselves by feedback effects, that is, by a process in which higher-level (systems) parameters influence lower-level components.

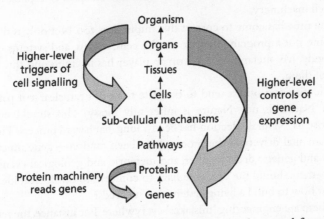

FIG 2. Figure 1 has been completed by adding the downward forms of causation, such as higher levels triggering cell signalling and gene expression. Note the downward-pointing arrow connecting from proteins to genes to indicate that it is protein machinery that reads and interprets gene coding. Loops of interacting downward and upward causation can be built between all levels of biological organisation.

Where is the program of life?

We have learned in previous chapters that the genome is sometimes described as a program that directs the creation and behaviour of all other biological processes in an organism. But this is not a fact. It is a metaphor. It is also an unrealistic and unhelpful one. The primary purpose of this chapter has been to steer us away from it. So we have focused on the cell machinery (proteins, membranes, organelles) that is necessary for gene transcription and for subsequent modification of the proteins thus produced. We have also developed the metaphor of the genome as an organ, which needs someone to play it. This can help us get our minds around the picture of biological function that we are starting to build up.

So far, so good. Still, the impression may remain that, okay, the genome is not a program, but that's not to say there *is* no program.

Rather, we just have to look elsewhere for it. Perhaps it is implicit in the cell machinery.

The time has come to correct that impression, too. Not only is the genome not a program, neither are the cells, organs, and systems of the body. My metaphor of the organ player has its limits and this is one of them.

Scientists and others tend to be quite fond of neat, clear-cut patterns. Nature is not. Nature is inherently messy. This should not surprise us. Natural selection has been a long, haphazard process. The fundamental drivers of the process have been random—gene mutations and genetic drift, weather, and meteoric and geological events. Why, then, should the outcome have to conform to our logical ideas about how to build a living system that succeeds?

There are engineering 'mistakes' everywhere. For instance, the retina of the mammalian eye is arranged backwards, so that light has to cross a tangle of neurons before it reaches the photoreceptors. And all over the body, nerves, vessels, and ducts follow strange pathways. In themselves, these make no sense. They are understandable only when we take into account the contingent evolutionary history. Nature has frequently gone down blind alleys from which it has escaped with strange results.

Living systems are simply not as they would be if all their component parts had been intentionally orchestrated. So when we talk of someone who 'plays' the genome, we must recognise that no one set of molecules is given a position of privilege over another. Nature always uses whatever comes to hand. That is also how it has evolved the pattern of regulation that we can to some extent imagine as the 'organist'.

The controlling networks of proteins, genes, membranes, organelles, etc. have all come together and continue to operate in quite a mixed-up fashion.

So for instance genes themselves can also be part of the regulatory networks that control gene expression levels. We will see examples of such networks in Chapter 5. Again, the components of a system may survive only because that higher-order system conforms to a particular, successful logic for survival of the organism, but that does not mean each component operates in an ideal way, or the best way in conformity with that logic. Indeed the system (the 'organist') has to accommodate all sorts of lower-level quirks to be viable at all.

The overall logic of the system may operate in a way that strongly resembles what happens when a program is running a computer, but that does not in itself indicate that there really is such a program. If we want to use a computer analogy, we could say that the regulation may simply be hard-wired, requiring no software. This points up the limitations of the computer model. Thinking in these terms, we seek to distinguish between the software program and the hardware that it controls. In living systems, there hardly seems to be any such distinction. Why should there be? There is no reason why nature should have developed separate software if it didn't need to.

The distinguished plant geneticist Enrico Coen put it well. 'Organisms', he said, 'are not simply manufactured according to a set of instructions. There is no easy way to separate instructions from the process of carrying them out, to distinguish plan from execution' (Coen 1999). Richard Dawkins expressed a related thought when he wrote:

If a computer is doing something clever and life-like, say playing chess, and we ask how it is doing it, we do not want to hear about transistors, we accept them. . . . We need *software explanations* of behaviour. I do not mean that animals necessarily work like computers. They may be very different. But just as the lowest level of explanation is not always the most appropriate for a computer, no more is it for an animal. Animals and computers are both so complex that something on the level of software explanation must be appropriate for both of them. (Dawkins 1976)

This expresses beautifully the multilevel nature of biological explanations that I am urging in this book. Significantly, this article was published in the same year as *The selfish gene*.

The 'organist', therefore, consists of regulatory networks of interactions at all levels, from the highest to the very lowest level, including networks incorporating genes themselves. There are no privileged components telling the rest what to do. There is rather a form of democracy, with every element at all levels having a chance to be part of a regulatory network. The co-ordinating hand is not so much an organist's as a conductor's. Or perhaps we should think rather of a 'virtual conductor'—the system behaves 'as if' it has a conductor. The genes behave 'as though' they are being 'played' by this conductor—rather like some orchestras that play without a separate conductor.

If we really want to use a computer analogy, we can. We can imagine a computer, which has been programmed to perform certain operations—but then that functionality has been hard-wired into the machine itself while the program has been lost. So it performs like a machine that is running a program, but it runs no program. In the case of a living system, however, the program does not and never did exist as such. We could perhaps view evolutionary history as a program (see Chapter 8), but that is really stretching it.

In sum, I hope this chapter has offered a strong antidote to the 'genes program everything' view. To that end, it has been useful to develop some alternative metaphors. But there are of course always limits to the validity of metaphor. They are ladders to understanding. When you have climbed them, you can throw them away.

5 § The Rhythm Section: the Heartbeat and other Rhythms

I think there is a world market for maybe five computers.

Thomas Watson (1874–1956), Chairman of IBM, 1943

Beginnings of biological computation

The world market in computers is today counted in billions. With the benefit of hindsight it is easy to laugh at Thomas Watson's prediction. I don't. I experienced at first hand the kinds of computer that he was referring to. Even 16 years later, in 1959, computers were still very few in number—massive, valve-based machines that took up vast space in university or industry basements and consuming so much power that, in post-war Britain, power itself was one of the factors limiting their use.

In the whole of London University there was only one such machine, called *Mercury*. It was guarded jealously like a god by the high priests of early scientific computing. Every second of time on the machine was precious. There was no question of wasting it on something as trivial as word-processing! Only the most serious number-crunching problems deserved this treatment.

If they were not computer scientists, the people who used *Mercury* were particle physicists, crystallographers, and numerical analysts. They knew their maths and had all learned how to program the computer via punched paper tape. They used machine code, and sometimes called in aid a protean version of what would eventually become one of the earliest computing languages to rise above the endless noughts and ones of machine code.

This was the context in which a young physiologist, not yet with a PhD to his name, and with little claim to any mathematical expertise, turned up on the doorstep of the Bloomsbury basement where the power-hungry god was housed to ask for some time to do his calculations. The answer was a firm 'No'. I was shown the door.

One of my professors had already suggested that it would be better to take my problem to Andrew Huxley in Cambridge. Together with Alan Hodgkin, he had already cracked the problem of calculating the electrical activity of a nerve. Later, in 1963, they were to win a Nobel Prize for this achievement. Andrew Huxley was already using another one of the earliest computers, a machine called *EDSAC* in Cambridge. But I was like a schoolboy with a precious bag of sweets: my first experimental results on the electrical activity of the heart. The idea that I should hand them all over to someone else, however distinguished, was not appealing.

However, this story is running ahead of itself. Why was I begging for time to do some mathematical modelling when most biologists were thought to be hopelessly innumerate? No other biologist in the whole university had dared to do so. And any claim I might have had was hard to support: I didn't even have an A level in mathematics, let alone a degree!

Reconstructing heart rhythm: the first attempt

I started my research career in physiology as a card-carrying reductionist. I was working on ion channels in excitable cells. The basic example of an excitable cell is a nerve cell or neuron. It's excitable in the sense that it can transmit an electrical impulse. Ions, as we have seen, are the key to how this happens.

It all starts in the cell membrane. This contains two layers of loosely bonded lipids with a space in between. One end of a phospholipid molecule repels water, the other attracts it; the inside of the cell is wet, and so is the environment in which the cell lives; so the lipids on the inside have their water-friendly ends pointing into the cell, and the outer level of lipids face the other way. The gap between them is liquid and labile (though not wet). Embedded in this lipid structure are proteins. It is by way of these proteins that interactions happen across the cell wall. Some of them actually stick out through the membrane at either end. Each end has a different molecular shape. Some shapes block and others facilitate the passage of ions through into the molecular structure of the protein. Once in, an ion will pass along that structure, all the way through and out the other side. And ions carry charges, remember. In this way, an electric current can flow through the wall of the nerve cell and so cause an electrical impulse to build up within it. This is how a nerve or a muscle is excited.

Anyway, in 1958 I was allocated as a research student to Otto Hutter, later Regius Professor at Glasgow. Otto had recently performed some heroic experiments with a German physiologist, Wolfgang Trautwein, on the way in which nervous activity can slow the heart rhythm via an action on cells in the pacemaker region—the region where the heartbeat originates.

Different proteins attract and channel different ions. In this case, we were dealing with potassium ions and the corresponding proteins, called potassium channels. So Otto set me to work on how potassium channels respond to electrical changes. In those days, before the full advent of molecular biology, this was about as reductionist as one could get. We now know that such channels are formed by proteins, and in most cases we also know which genes code for these proteins.

Otto and I had an early and important success. We were building on the work that Hodgkin and Huxley had done on nerves. They had shown that potassium channels open up during each nerve impulse. We therefore expected to find these kinds of channel in the muscle of the heart, too; and we did. We found that in the heart they are much slower. That also was expected. An electrical impulse in a nerve cell lasts only a thousandth of a second. In the heart, impulses last hundreds of times longer—for a large fraction of a second. The heart rate in a human is around 60 beats per minute so there is a whole second for the impulse to occur and for the heart to then recover before the next beat.

Then came the big surprise. We found some potassium channels that rapidly *close* during each heartbeat. They do just the opposite of what we expected. This is the real joy of experimental science. When something new turns up you can't contain yourself. You want to rush out and tell the world about it.

But could there be something wrong with the experiments? Actually, we didn't seriously think that was the case, but we did have reason to be careful. Report an unexpected discovery and you must expect your peers to be critical. So, we performed the obvious control experiments to rule out some of the possible snags. Moreover, Bernard Katz (also to become a Nobel Prize winner for his work on neuromuscular transmission) had found similar channels in other kinds of muscle in the body. So, the heart was not the only organ in the body where these channels were found.

It also dawned on us that such a process could form part of an effective energy-saving mechanism. It all has to do with ion gradients. If there are more ions of a certain sort outside a nerve or muscle cell than inside it, they tend to migrate inwards; and vice versa. That is, ions diffuse down their electrochemical gradients, which lead either outwards from the cell (in the case of potassium ions) or in towards it (in the case of sodium ions).

Now consider the muscle cells that make up the heart. When the heart beats, a strong electrical impulse passes through these cells. That is, a mass of charged ions get shunted inwards or outwards through each protein channel. So far, so good; but you can have too much of a good thing. If too many ions move, then, before the heart could beat again, it would be necessary to do a lot of work to push the ions back again—to restore the ion gradients.

So it made sense for those apparently anomalous potassium channels to close during the heartbeat. This helped to conserve the ion gradients and so to economise on energy expenditure over time. The effect is large. Greatly reducing ion movements during each heartbeat reduces the total energy required to restore these movements and so maintain the activity. The change in energy requirement is around ten-fold.

This was my first encounter with integrative systems biology. It drew my attention to the overall logic of a higher-level biological process and how that can help to explain why a system works as it does at lower levels and how it has evolved that way. I could also see that this sort of effect could happen on many levels. Yes, it was possible to view the mechanism as energy-saving; but also, the two patterns of potassium channel behaviour could combine with the action of other known ion channels in the heart to generate rhythmic activity.

The experimental encounter with natural reality was starting to teach me a new kind of thinking. At the time, of course, I did not fully see that or appreciate how significant this kind of thinking was to become 40 years later. Instead, I began reaching out for one of the plums of cardiac physiology: how to account for the heart's pacemaker rhythm. Suddenly it seemed within reach.

The idea was to explain this rhythm in terms of the complex behaviour of the various ion channels. Others had already proposed this approach; in broad terms, the theory itself wasn't novel. What were new were the experimental data. Could we perhaps generate sufficient data to support the theory *quantitatively*? That would require mathematical analysis and the construction of a mathematical model, but the possibility was tantalising.

As a student, I had been riveted by the way Hodgkin and Huxley had, just a few years earlier, modelled the conduction of nerve impulses. Their paper, 44 pages long and peppered with mathematics, is a monumental achievement. As an undergraduate I did not fully understand it, but was greatly impressed. This was how biology could become quantitative, just like the physical sciences.

I had heard that Andrew Huxley had spent six months working with a hand calculator (a crunching German machine called a *Brunsviga*) to do the calculations. Could something similar be done for the heart? And would it be possible in this way to reconstruct pacemaker rhythm? I even had a *Brunsviga* to play with.

But it took months with a *Brunsviga* calculator for just a few thousandths of a second of nerve activity. How much longer was it going to take to calculate heart activity for a whole second? Clearly, an electronic computer was necessary. Calculations that would take years were not an option for a PhD student in need of a thesis.

That, then, is how I came to take my experimental results and a few hand-waving feasibility calculations along to the guardians of the

Mercury computer. I must have explained, breathlessly, and in mathematically quite naïve terms, what I hoped to achieve. I would fit my data to some non-linear expressions, I thought, then would solve the differential equations for the electrical state of the cardiac cell—and then, (hey, presto!), I would see oscillations emerge from the computer's output.

A single question stopped me in my tracks. 'Mr Noble, where is the oscillator in your equations? What is it that you expect to drive the rhythm?'

I was speechless. I had no idea how to reply.

On a sheet of paper, I sketched out the physiological interactions that I thought would work. This served only to confirm my mathematical naïvety. There was no oscillator function in my equations. Thirty years later, and after many further rounds of mathematical modelling of cardiac rhythm, I was asked almost exactly the same question by the science correspondent of a national newspaper. By that stage I knew the answer: *Silly question!* (Of course, I did not put it quite that way to the journalist!)

Yet, it looks like a reasonable question. In a system that oscillates, it seems that there must be some specific component that oscillates, around which the behaviour of the entire system is geared; and there must be a mathematical function that describes the way that component oscillates. Indeed, it is an eminently necessary question, if we are talking about some man-made, mechanical systems. But we are not. Instead, we can have a system that operates rhythmically and yet contains no specific 'oscillator' component. There is no need for one. The reason is that the rhythm is an integrative activity that emerges as a result of the interactions of a number of protein (channel) mechanisms.

So there is no need for any oscillatory equation at the molecular level. The rhythm is a systems property. Some biologists have called these properties 'emergent' properties. I prefer 'systems-level' properties, but we are talking about the same kind of phenomenon. Still, even if I had known the answer way back when I was discussing my project with the guardians of the *Mercury* computer, I am not at all sure I would have convinced them. They also were committed reductionists. If rhythm was to emerge from the calculations, they assumed there must be a rhythm generator. They wanted to see at least one mathematical equation with a sine wave oscillation or similar function. There wasn't one; so they weren't buying.

In the end, however, I wore them down. I spent some time attending maths lectures for the engineering students. I started to learn how to handle matrices, differential equations, Bessel functions, complex numbers, you name it. I even picked up a manual and started to learn the gibberish programming language needed to convert my equations into computer code that would go onto rolls of paper tape. Then I went back and knocked on the basement door again.

Reluctantly, they agreed to give me a chance. It was calculated that I would need around two hours per day and these I could have . . . between 2 and 4 in the morning!!

So my research day started at 1.30 a.m.: a quick coffee, and then two hours at the *Mercury* computer. Then on to the slaughterhouse at 5 a.m. to pick up the sheep hearts with which the day's experiments would be done. Those experiments sometimes lasted until the time came to return to programming *Mercury*. I think that experience completely wrecked my circadian rhythms, but let's return to that kind of rhythm later in this chapter.

The integrative level of heart rhythm

The story for this chapter is a personal one. But it relates strongly to the theme of this little book. I was lucky in all respects. Otto Hutter and I got some unexpected and important experimental results. I learnt enough maths and computer programming to put it all together into a model. And a few months later there were two papers in *Nature*. The idea worked.

There have been many refinements of that early work since then. The cellular models have become much more complicated. And they have been incorporated into impressive whole-organ anatomical models created by colleagues working at the University of Auckland in New Zealand to create the first example of a virtual organ, the virtual heart. But the insights of the early models remain broadly correct. In brief, it suffices to take account of the activities of fewer than 50 proteins to account for the rhythmic activity of cardiac cells. This is a good example of nature using modular systems.

The basic rhythmic mechanism is generated by a relatively compact and tight-knit network of proteins and the genes that code for them. It only works in the context of what is happening elsewhere in the body—the activity of many other proteins. But in that context this little modular system is sufficient to produce what I think you'll agree is quite a striking effect.

Let's have a look at how the integrative activity works in this case. The point of the story is that causation works both ways, up and down. The components alter the behaviour of the system, and then the system in turn alters the behaviour of the components. The system is the cell—a muscle cell in the heart (in this case a rabbit heart pacemaker cell—Fig. 3). The components are the protein molecules that channel the electrically charged ions (of, in this case, first potassium, then calcium, and finally a mixture of elements).

The rhythmic behaviour is shown by the voltage or electrical potential. As the heart beats, the voltage in the relevant muscle cells goes up and down. So does the flow of ions through the protein channels. These two patterns of oscillation are clearly linked. How?

Well, for one thing, the operation of the protein channels drives the critical rhythmic activity of the cell. There is no voltage change unless ions flow through the channels. That is obvious, and that is where we start. The point of the story is, it works the other way round, too. The rhythmic activity of the cell drives the operation of the protein channels. We can show this by turning off the feedback from the cellular rhythms to the proteins that channel the ions. If we do that, the system as a whole ceases to function. Neither at the cellular nor at the molecular level is there any longer any sign of oscillation.

Internally within the protein channels, nothing has changed. So if this biological phenomenon were produced exclusively by bottom-up causation, there is no reason why it should stop at this point. But it does. Clearly, therefore, downward causation, that is, the feedback effects from the system to the components, is essential for the system to function.

Of course, it is not so easy to turn off the feedback effects in an actual living heart. What we can do is to develop a computer model. First we identify what is happening in isolated pacemaker cells, measuring and intensively analysing the results. Then, on that basis, we develop a computer program that performs a continuing stream of complex sets of calculations—and the results of those calculations match the results we get by measuring the actual biological phenomenon.

As the cellular voltage oscillates and the cell pulses, the environment of the protein channels changes; and that influences their behaviour. Accordingly, for our model to work, we have to build in some calculations which represent the impact of the changing cellular environment on the protein molecules.

Our computer program runs those calculations, along with all the other calculations that represent the behaviour of the proteins and the ions. It produces the correct results—it successfully models the behaviour of the living heart. Then, after a while, we change the way the program runs, so that it continues to perform all the other calculations but not those which relate to the feedback effects. Immediately, the system as a whole, as represented by the computer model, seizes up.

Figure 3 shows a picture of this.

The top trace shows how the cell voltage varies with time. The bottom traces show how three of the protein channel mechanisms vary with time. These include a potassium channel, a calcium channel, and a channel that carries a mixture of ions. There are more proteins involved in the model than I show here, but if I included all of them the figure would become very confusing. The horizontal axis is measured off in milliseconds (so there are 1000 in one second). The vertical axis shows millivolts for the voltage trace and nanoamps for the current traces representing the activities of the protein channels.

In the first second (1000 milliseconds) there are four oscillations of cell voltage, and corresponding oscillations of protein channels. After four beats, the feedback from the cell voltage to the protein channels is cut by holding the cell voltage constant. If one or more of these oscillations of the protein channels were driving the cell voltage then

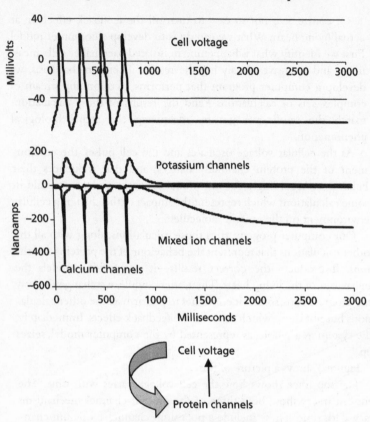

FIG 3. Computer model of pacemaker rhythm in the heart (Noble and Noble 1984). For the first four beats the model is allowed to run normally and generates rhythm closely similar to a real heart. Then the feedback from cell voltage to protein channels is interrupted. All the protein channel oscillations then cease. They slowly change to steady constant values. The diagram shows the causal loop involved. Protein channels carry current that changes the cell voltage (*upward arrow*), while the cell voltage changes the protein channels (*downward arrow*). In the model, this downward arrow was broken.

they would continue to oscillate on their own. But this is not the case. The oscillations of the protein channels cease. In each case, the line that shows their levels of activity flattens out. Clearly, the feedback from the cell voltage to the protein channels is an integral part of the rhythm generator.

One of the protein channels in Fig. 3, the mixed ion channel, seems to do rather little until the voltage feedback is interrupted, when it then becomes very large. We will return to this channel protein in Chapter 8.

Systems biology is not 'vitalism' in disguise

Reductionist science is often presented as 'hard' science, 'real' science, while integrative, systems-level science is sometimes viewed as 'fuzzy'. One reason for this prejudice is historical. Biological science had to struggle to emerge from the days of vitalism, when people thought that something non-physical had to be added to matter for there to be life.

But there is nothing 'fuzzy' about the example I have discussed in this chapter. On the contrary the ideas have been encapsulated in the same way as the hardest of reductionist explanations, that is, in the form of mathematical equations with particular solutions.

Nor is it reductionism in disguise

Perhaps this is the reason why some would claim that this is just reductionism in disguise. They would say, correctly, that the successful simulation depends on knowing the relevant properties of all the molecular components. From there on, the reconstruction is 'bottom-up'; the rest follows. What, they would go on to ask, is the essential difference between this and a completely reductionist explanation, for example one in which the oscillator had been found to be a molecular one which then drives the cell, which is what the guardians of *Mercury* were looking for in my equations? As we will see later in this chapter, such biological oscillators exist.

The answer is that the mechanism necessarily involves feedback from a high-level property, the cell electrical potential, which determines the activity of lower-level mechanisms, the individual ion channels. This is what distinguishes an integrative explanation of a biological phenomenon. There is downward causation.

To be sure, successful reduction is also involved. In this sense, reduction and integration are two sides of the same coin. But to claim that this is thereby a reductionist explanation would be to deprive the term 'reductionism' of any meaning. We would simply be saying that 'scientific' means 'reductionist' because 'reductionist' means 'scientific'. And then we would need to invent two more words to distinguish between explanations that are completely 'bottom-up', with causation running in one direction only, from those that also involve causation running in the opposite direction.

There is an interesting asymmetry between reductionists and modern integrationists in biological science. An integrationist, using rigorous systems-level analysis, does not need or wish to deny the power of successful reduction. Indeed he uses that power as part of his successful integration. Many reductionists, by contrast, seem for some reason to require intellectual hegemony.

Perhaps the problem is that for some scientists reductionism functions as a security blanket. It avoids the need to ask too many questions, to stare into the abyss of fundamental uncertainty. If we abandoned the universality of the reductionist approach, who knows what would happen? For sure, the nature of biological science would change. But so it should!

This change is long overdue. We need to take on board the reality that causation and explanation do not always run upwards from lower to higher levels. We have many dramatic examples now to show that causation can run in the opposite direction. Systems effects can even control the ultimate low-level processes of gene expression (Chapter 4). So how can anyone possibly imagine that we will progress without recognising the need for integrationist explanations to complement the reductionist at all levels?

Cardiac rhythm and arrhythmias are beautiful test cases for these ideas. We have seen that the normal rhythm generator mechanism is integrated at the cell level. In addition, there are abnormal rhythm generators at levels below and above the cell level. If biological science is to deliver on its promise of continuing improvement in human healthcare, it must apply the integrationist perspective here, too.

Consider the life-threatening abnormal rhythms that we call tachycardias (very rapid beating) and fibrillation (desynchronised beating leading the heart into a kind of tremor). These pathological conditions at the multicellular level correspond to similarly striking subcellular phenomena. What happens is that the calcium signalling system falls into oscillatory patterns. How to understand what is going on here? Clearly it is necessary to simulate electrical activity at the level of the whole organ. It is true that we need detailed knowledge of the protein channel mechanisms involved here. But we cannot therefore say that all of these mechanisms are molecular. That would be to blur critically important distinctions, ones that can mean the difference between life and death.

Fibrillations kill. We don't understand that by studying genes and proteins alone. We don't even understand it by studying cells alone. Such a phenomenon becomes explicable only in terms of the way in which large numbers of cells interact at the level of the whole organ.

Some of us may be happier with nice, neat detached molecular studies, and there is nothing wrong in that. But let us not therefore subscribe to the convenient ideology that says there is no point in investing time and money at other levels of biological science.

I conclude this section by noting that, for presentational purposes, I have simplified the analysis of the heart pacemaker rhythm. There are many other processes that are involved, particularly in fine-tuning the process and in ensuring that it is robust. We will return to the issue of robustness in Chapter 8.

Other natural rhythms

Heart rhythm is the most obvious of biological oscillators. We all feel our pulse rhythm and doctors use it for diagnosis. It is remarkably regular; when Galileo was discovering that the length of pendulum determined the frequency of its oscillation, he used his pulse to time his observations. It is no coincidence that we use the word 'beat' both for heart rhythm and for musical rhythms. Early musicians almost certainly based their rhythms on the approximately 1 hertz frequency of the heartbeat; then they broke that down into shorter components to obtain higher frequency rhythms, and counted several 'beats' to obtain lower frequencies.

But nature is full of other rhythmic mechanisms, varying over an astonishing range of frequencies from over 1000 per second to just once every few years. Thus, many nerve cells generate rhythmic impulses at frequencies much higher than that of the heart. Since a nerve impulse lasts around 1 millisecond, the highest frequency in this case approaches 1000 per second. Also at a frequency higher than that of the heart, the wings of flying insects oscillate around 200 times a second.

The rhythm of respiration is the most obvious one at the next level of frequency, a little below that of the heart—once every few seconds. Hunger and thirst rhythms come next at several times per day. Then there are circadian rhythms determining roughly 24-hour cycles of biological activity (such as sleep and waking), lunar-based (circalunal) rhythms such as human ovulation, annual (circannual) rhythms determining, for example, animal migrations, and the multi-annual rhythms determining reproduction and social behaviour over periods of more than 10 years in some species, including cicadas that follow a rhythm with a period as long as 17 years.

The oscillation period of a biological rhythm can therefore vary from around 1/1000th of a second to around 20 years, a range of around a trillion-fold.

How similar are all these in their properties to that of heart rhythm? Is there a basic pattern of natural biological oscillations or must we work each case out individually? The answer is that they are nearly all different. Other than the need for delayed negative feedback loops to generate the rhythms, they represent almost all levels of biological integration from the molecular to whole organs and systems. At the higher frequencies, neuronal rhythms share with the heart's pacemaker the property of integration at a cellular level, with a number of proteins forming ionic channels whose kinetics are determined by the cell electrical potential.

Respiratory rhythm is more like the higher forms of cardiac arrhythmia in requiring the co-operation of many nerve cells, peripheral and central, organised in the form of a feedback system. One might imagine, therefore, that the even slower rhythms, circadian, circalunal, circannual, and multi-annual, would also most naturally find their level of integration at a supracellular level.

In some senses this is true. Large numbers of cells are involved. What neuroscientists call the suprachiasmatic nuclei (SCN) is the group of cells in the mammalian brain that are necessary for circadian rhythm. It is around 200 000 cells in magnitude (that is about the same number of cells as in the heart's pacemaker region). Feedback loops between various parts of the organism—neuronal, hormonal, and cellular—are also involved.

Yet, over the last two or three decades, a surprising insight has developed. It has been shown with spectacular success that the fundamental mechanism inside the cells of the SCN is molecular. The feedback loops involve daily rhythms in gene expression level, generated by further feedback loops involving particular protein and gene components. Even single cells isolated from within the SCN can show these circadian rhythms in gene expression.

Single cells with 'molecular' clocks! Perhaps here, at least, we have an example of a completely reductionist account of a major biological phenomenon. At first sight this seems to be so.

Mutations in a single gene (now called the *period* gene) are sufficient to change the circadian period of fruit flies (Konopka and Benzer 1971). This discovery of the first 'clock gene' was a landmark, since it was the first time that a single gene had been identified as playing such a key regulatory role in a high-level biological rhythm. Does it disprove the general thesis of this book that genes must always act in co-operation to generate higher-level functions? There are two ways of answering this question.

First, regulatory mechanisms are often complicated. So if a single key component is damaged, the whole mechanism is apt to malfunction, perhaps in ways that are unpredictable. Most mutations, after all, are a form of damage. Accordingly, when the absence or aberrance of one component is shown to stop a system working, we have to be careful how we interpret this. When a child tinkers with a toy until it ceases to work, it is not necessarily the last thing she did that broke it. The adult version of this scenario is rather different: we wait until the TV set goes wrong of its own accord and then see whether shaking the box of electronics kicks it back into order! When that works, often enough the problem was a single bad connection.

In biology, some gene mutation and knockout experiments are like that. The results are difficult to interpret. That said, the case of the *period* gene in the fruit fly is different. The expression levels of this gene are clearly part of the rhythm generator. They vary (in a daily cycle) in advance of the variations in the protein that they code for.

But more than that is going on here. The protein is involved in a negative feedback loop with the gene that codes for it (Hardin *et al*. 1990). The idea is very simple. The protein levels build up in the cell as the *period* gene is read to produce more protein. The protein then diffuses into the nucleus where it inhibits further production of itself by binding to the promoter part of the gene sequence. With a time delay, the protein production falls off and the inhibition is removed so that the whole cycle can start again. So, we not only have a single gene capable of regulating the biological clockwork that generates circadian rhythm, it is also itself a key component in the feedback loop that forms the rhythm generator.

We call such causal loops feedback loops because they form self-contained regulatory processes. Feedback is involved in the great majority of cases where downward and upward causation are linked together in biological systems. The linkage is a defining feature of feedback since the downward causation modifies the components responsible for the upward causation, which in turn modifies those generating downward causation . . . and so on. Mathematical models of such loops are easy to construct and they are robust and explanatory.

The second way of answering the question is more subtle. The basic rhythm generator in this case does seem to be dependent on a single gene and the protein it codes for. But does it carry out its work in isolation? Is it a 'single gene module'? The answer is a resounding 'no'. The further researchers get in unravelling the molecular feedbacks involved in circadian rhythms, the more gene and protein components appear to be involved.

This has emerged clearly from studies of circadian mechanisms in other animals such as the mouse. Moreover, these rhythmic mechanisms do not work in isolation. There has to be some connection with light-sensitive receptors (including the eyes). Only then will the mechanism lock on to a proper 24-hour cycle rather than free-running at say 23 or 25 hours. That is why, as Foster and Kreitzman have written: 'What we see as a clock may well be akin to an emergent property of the system and all the genetic paraphernalia merely part of the fine tuning' (Foster and Kreitzman 2004).

So, yes, it is impressive that a particular gene can be regulated by negative feedback from the very same protein it codes for. Nonetheless, we don't actually have here an example of a 'single gene function'. But there is a more important point still. Suppose that the original simple feedback between this *period* gene and the protein it codes for had indeed been all that was required. Even then it would still not have qualified as a 'single gene function'. The reasons are critical to the message of this book so let me spell them out again.

First, the protein that is coded by *period* is like all molecular components—it operates within the context of the complete cell. It depends for its production on the transcription/translation mechanisms and the ribosome machinery. Its ability to access the promoter region of the *period* gene depends on properties of the nuclear membrane.

It is not only genes that never operate outside the cellular context. The same applies to individual proteins. It is conceptually convenient to 'isolate' the *period*–protein system, sure. This does indeed help us to appreciate its unusual characteristics as a molecular level oscillator. But this is an artificial conceptualisation. The real living system operates only in the context of the functioning of many other genes and proteins.

Second, why do we call this the *period* gene? Because it codes for the protein that, in feedback with the gene, generates a periodic function. Well, that is the first function of this gene that we have identified, sure enough. But how do we know what other functions it is involved in?

Again, listen to what Foster and Kreitzman have to say: 'what we call a clock gene may have an important function within the system, but it could be involved in other systems as well. Without a complete picture of all the components and their interactions, it is impossible to tell what is part of an oscillator generating rhythmicity, what is part of an input, and what is part of an output. In a phrase, it ain't that simple!' (Foster and Kreitzman 2004: 120).

Indeed not. The *period* gene has also been found to be implicated in embryonic development as the adult fly is formed over several days. And it is deeply involved in coding for the male love songs generated by wing-beat oscillations which are specific to each of around 5000 species of fruit fly and ensure that courtship is with the right species.

Fruit fly amorous troubadours! The music of life has some great surprises!

Period is therefore rather like one of those very useful pieces of Lego which enable a child to build many different kinds of structure. This is one of the great problems of gene ontology, and one of the reasons for preferring relatively neutral labelling of genes. Labelling the sodium–calcium exchanger as *ncx* is a good example. It tells us no more than the protein-level function, which is to exchange sodium and calcium ions (Na–Ca-exchange → *ncx*). It does not pretend to tell us what role such exchange plays in, say, heart rhythm, or vision. Such modesty is not at all out of place. When, instead, we choose to label a gene as, say, a 'clock gene', we may well be blinding ourselves to what else it does.

Recall the *Dscam* gene referred to in Chapters 1 and 2? I forgot to explain its name then, but now is a good time to reveal it. It is the gene coding for a protein called the Down's Syndrome Cell Adhesion Molecule, from which it is clear that it was first annotated functionally as important in Down's Syndrome, a birth defect causing mental retardation. It is now thought also to be involved in an extensive insect immunity system.

We will return to this kind of problem of gene re-use, and variation of functionality between species, when we consider the role of evolution and the modularity of nature in Chapter 8.

6 § The Orchestra: Organs and Systems of the Body

> I know one approach that will fail, which is to start with genes, make proteins from them and to try to build things bottom-up.
>
> Sydney Brenner, 2001

Novartis Foundation debates

This chapter also begins with a personal story.

The Novartis Foundation is a unique institution. Although based in the UK, it was founded originally by the Ciba Company of Basel. Ciba became Ciba-Geigy and then merged with Sandoz to become the pharmaceutical giant Novartis. The Foundation organises discussions on biological themes. Only a few people are invited. The proceedings are recorded, transcribed, and then published in their entirety. Read the volumes and you feel as if you are listening in to the live debates.

I have been privileged to take part in three seminal Novartis discussions on the nature of biological science (Novartis_Foundation 1998; Novartis_Foundation 2001; Novartis_Foundation 2002). Some of the ideas in this book were honed in those debates—and I have also freely borrowed from some of my fellow participants. One such is Sydney Brenner. Sydney is one of Britain's most distinguished molecular geneticists, and richly deserved his Nobel Prize in 2002. He works at a very different level from me, which makes his insights into linking levels in biology all the more pleasing.

At the first of these meetings I played a game with the other participants. I presented my work on heart rhythm (Chapter 5) and challenged them to say what sort of work this was. In particular, was it reductionist or integrationist? The answers split roughly 50–50! At the second meeting I tried to be a little more helpful and say what, in my view, made this an example of integrative systems biology.

To see what makes this result interesting, we must understand that there is a well-worn path in such debates. The procedure is to contrast two opposite ways of simulating and understanding complex biological processes. One is the bottom-up approach. This, as we have seen, starts with genes, reconstructs protein sequences and structure, then goes on to protein function, followed by the biochemical pathways that the proteins form. The reconstruction process continues on, up through all the biological levels (Fig. 4), until eventually one reaches organs and systems, and, hopefully, the whole organism.

The alternative is what people commonly refer to as the top-down approach (using that term a little differently from the way we have done so far in this book). This is the approach of classical physiology. It starts with the overall behaviour of systems. It analyses the circulatory system, the respiratory, the immunological, neuronal, and reproductive, and so on. Then it progressively identifies and explores the elements of each system so as to deduce the underlying functions and mechanisms.

The first approach has serious problems. The second doesn't completely avoid them.

Problems with bottom-up

The bottom-up approach suffers from two fatal problems. The first is computability.

Consider the problem of folding a protein. It is the three-dimensional structure of a protein that makes it work the way it does, and that structure reflects the way the protein is folded. The problem is to reconstruct the chemical processes involved in the folding of a single protein. This may not seem like a huge problem—just one protein, just a few folds. But suppose we give it to the most powerful computer in the world—a mammoth called *Blue Gene*, which IBM is

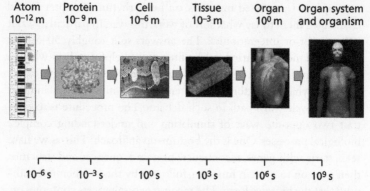

FIG 4. Size and time scales involved in the various levels of organisation of the human body. Between atoms and organisms there is a range of 10^{12} in size (from a trillionth of a metre to around a metre) and 10^{15} in time scale (from microseconds to tens of years) of major structures and events. (From Hunter *et al.* 2002, reproduced with permission.)

constructing at a cost of \$100 million. Suppose *Blue Gene* devotes its full calculating power to solving the problem. How long will it take? Probably months!

The molecule we are talking about is around one nanometre in size—one billionth of a metre. The chemical processes involved take about a millionth of a second to complete. Compare that with a cell, which is typically tens of micrometres in size (several millionths of a metre).

For an exhaustive molecular reconstruction of the activity of a single cell, we would need to simulate the interactions of around 10^{12} such molecules. And we would have to continue the simulation many seconds, minutes, hours, days, or even years, a span of time scales of around 10^{15}. This would require unimaginably large computational resources—around 10^{27} *Blue Genes*. There simply won't be enough stuff in the whole solar system to build such monsters. Yet, this would be only the beginning of our problems. The project is to reconstruct the tissues, organs, and systems of the body, remember. Even the smallest of organs in the body has millions of cells.

You will have got the point. I don't need to labour it any further. This approach fails on the first hurdle. It is unbelievably impractical.

Some would say 'No matter; the principle is right. Man is nevertheless just a mass of molecules.[1] In principle it must be possible to *imagine* such a complete bottom-up reconstruction, even if we, and any likely successors, are never going to have the computing resources to make it possible.'

This brings us to the second fatal problem with this approach. Let's grant the imagination proposed. What would it achieve? It would allow us to show that our understanding of fundamental molecular processes is correct in living organisms, as well as in the inorganic realm, and that at that level no other processes are necessary. That is the most it could achieve for, as we have seen, the characteristically biological systems-level explanations would elude it. They don't exist at the molecular level.

So under what circumstances would that finding be relevant? Who does not believe that the laws of physics and chemistry govern the molecular events in a living system? Where are the vitalists that we need to convince? I see none amongst my modern biological science colleagues. Probably, the last remnants of any form of vitalism were in the generation of neuroscientists like Sherrington and Eccles, in the early and mid-twentieth century, who favoured a dualistic understanding of the brain. I will deal with that idea in Chapter 9.

But the debate between systems biology and reductionist biology has got nothing whatever to do with the debates on vitalism or dualism. As I showed in Chapter 5, integrative systems biology is just as rigorous and quantitative as reductionist molecular biology. The only difference is that it accepts that causality goes from higher to lower levels as well as upwards.

We don't need convincing that, at the molecular level, molecular processes are all there are. But, to reiterate the theme of this book, that does not mean that we are nothing but a bunch of molecules.

---

[1] 'You, your joys and your sorrows, your memories and your ambitions, your sense of personal identity and free will, are in fact no more than the behaviour of a vast assembly of nerve cells and their associated molecules' (Crick 1994).

Structures and processes at a higher level simply are not visible at the molecular level. The genes and proteins of the body do not in some way 'know' or 'reveal' what they are doing in higher-level functions. The assumption that they do is a strange one. We might indeed be forgiven for suggesting that it is itself a superstition of the order of vitalism.

While vitalism no longer finds favour amongst biologists, it is very much alive and well elsewhere in society. I would like to convince some of my readers in this camp that what they may be looking for in seeing life as a whole rather than the sum of its component parts may well be satisfied by the systems biological approach. Life is wonderful enough. We don't need to endow it with mystery to appreciate that.

Problems with top-down

The top-down approach has been successful as a tool in physiology for many years. Take our understanding of the circulation. Most of the knowledge that is used in medical practice has been derived from a systems-level approach that works by identifying the components that need further study at progressively lower levels. In this way, physiology has succeeded in 'drilling down' to lower and lower levels. Indeed, molecular biology itself represents the limiting term of this progression.

Consider the transport of oxygen in the blood. This was identified as depending upon the red blood cells, and then on a molecule called haemoglobin located in these cells. Finally we unravelled the molecular biology involved in the interaction of oxygen with haemoglobin. This process has been used time and time again. It is the basis of the great success story that has been reductionist biology in the twentieth century.

So what is the problem with this approach? It is that, having burrowed down through the levels to the smallest components, the molecules, we now need to build the whole thing up again *quantitatively* if we are to understand it at a systems level. The problem is that this brings us straight back to the difficulties of 'bottom-up' reconstruction. Understanding the components is necessary but not sufficient for systems-level understanding. Reductionist analysis is only half the story.

Middle-out!

There is now a discipline of quantitative computational physiology. Despite all the difficulties, this does succeed in reconstructing organs and systems at many different levels. Later in this chapter, I will describe the 'virtual heart' that has been reconstructed using a combination of the computer modelling of the cell, as described in Chapter 5, and impressively detailed anatomical and mechanical models of the whole organ.

Once, while presenting this 'virtual heart' to a Novartis discussion, I tried to capture the essence of what we had done. I wanted to explain the reasons for what has been at least a partial success in integrating between three major biological levels—protein channels, cells, and the whole organ.

At this point Sydney Brenner simply remarked 'Middle-out! What you are doing, Denis, is neither bottom-up nor top-down. It is middle-out!'

That is Sydney Brenner all over. One minute you think he is half asleep and no longer following the debate; the next minute a logical stiletto has suddenly penetrated the fog of the discussion. What did he mean? The concept is simple and pragmatic (Fig. 5). Biological function happens at different levels. We can gather quantitative data at any level. Once we have enough of it to feed into a simulation, we can start a systems analysis at that level.

All levels can be the starting point for a causal chain, so any of them can be the starting point for successful simulation. In networks of interactions at many levels, there is in fact no alternative. Analysis must start somewhere, but it doesn't really matter where. It might be at the level of gene–protein networks, or cell function, or organ structure. This is the 'middle' part of Brenner's metaphor. There can be many 'middles'. Brenner's would be different from mine. I start with cells, he starts with genes. That doesn't matter. In the best of all systems-biological worlds, we will all eventually meet up, anyway.

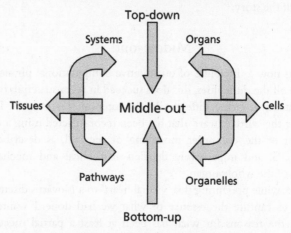

FIG 5. Relationship between the bottom–up, top–down, and middle–out approaches to reconstructing the organs and systems of the body. This diagram envisages the middle level being cells or tissues, but any biological level can be a middle starting point. The essential concept is that, since all levels can be the starting point for a causal chain, any can be the starting point for successful simulation. There is no privileged level in biological systems that 'dictates' the rest. (From Noble 2002, reproduced with permission.)

Then, when we have established sufficient understanding and success at our chosen level, we can reach out (this is the 'out' part of the metaphor) to other levels. Ideally, we eventually reach right down to the level of genes, and right up to the level of the organism. To the degree we can interpret the genome in terms of physiological function, this is how we will do it. Linking levels is part of what systems biology is about.

The critical reader may complain, 'How on earth does this differ from starting at the bottom in the first place? This must be a cheat! Surely, the approach will eventually, perhaps very quickly, fall foul of the same fatal problems?' No, it doesn't.

The reason is that this way we can select what we are interested in. As we reach down towards lower levels, we can rely on our higher-level analysis to identify just those features of the lower-level mechanisms that are relevant, and we can ignore the rest. The lower levels are seen through the filter of the higher level. This allows us to highlight what is important in the otherwise overwhelming mass of data. It greatly reduces the amount of information that we must carry over from one level of analysis to the other.

Here is an example from my area of cardiac physiology. Some people have genetic mutations that alter the electric charge on parts of the protein the affected genes code for. This leads to changes in protein function. It is now possible to identify what is important about these changes. That is, we can carry our analysis forward up the levels right through to the whole organ. In this way we can understand how these genetic mutations can cause sudden heart death.

We can do this because we start with the systems-level phenomenon, the cardiac arrest. That is what we have to explain. That imperative guides our investigations into what is going on at the lower levels. If we restricted ourselves to the 'gene's-eye view' we would never achieve comparable results. We could deduce that there is an electrical charge change in the protein that the gene codes for, but could never determine whether that would be sufficient to cause a fatal heart arrest. That depends on the higher-level events.

A similar economy of understanding and computation applies when the middle-out approach is used to reach out to higher levels. We don't need to incorporate all the details of the lower levels. Instead we identify those features that are functionally important at the higher level.

In the physical sciences, there is a discipline that has used this approach for many years. It is called engineering. An engineer also selects the level and detail of simulation he needs according to the problem he is tackling. It is not necessary to understand all the molecules in order to model and construct a bridge, for example.

Engineers also use the principle of modularity. The latest Intel processor inside your laptop computer may have as many as 200 million transistors. Who can understand how it all works? No-one! No-one needs to. If those who make each module understand what they are doing, and what they produce fits into the network as a whole, that is sufficient. This is not a bad analogy for the way organisms are put together during development. The individual parts do what they do without knowledge of the whole system (Coen 1999).

The organs of the body

Clearly, the orchestra of life is large. How large? How many sections does it have?

There are about 200 distinguishable types of cell in the body. We could regard these as the individual musicians. They are organised into several organs and systems. Amongst the organs, we have the brain, the heart, the liver, the kidneys, the pancreas, the stomach, the lungs, the genitals, and various endocrine glands. These are further organised into systems, such as the nervous system, muscular–skeletal system, circulatory system, respiratory system, endocrine system, immune system, reproductive system.

So, just as a full-scale musical orchestra has its strings, woodwind, brass, keyboard, choral, and rhythm sections, so this orchestra has around a dozen sections, the organs, organised into five or six groups, the systems. It is large enough to play life's equivalent of Beethoven's ninth—the all-singing, all-dancing human being.

The aim of physiology is to understand how it works. How well are we doing? As a physiologist myself, I am no doubt biased. But I think it is fair to say that, at many levels, the job has been very well done. We know the functions of all the organs and systems, and how they interact. This is the basic science that informs medical practice today.

At some levels, particularly molecular and cellular, we now also have a good quantitative understanding of key processes, such as how a muscle contracts, how a nerve transmits information, or how the pancreas secretes insulin. Some of these processes are understood in remarkably fine detail. At one of the Novartis Foundation meetings, after K.C. Holmes from Heidelberg had described the molecular events in muscle contraction, the chairman, Lewis Wolpert, asked

'How much more do you want to know? It seems to me that you have already solved the problem' (Novartis_Foundation 1998).

Holmes rapidly pointed to where there are still gaps in our knowledge. But Wolpert's basic reaction was sound. At the molecular level, and in certain other areas of physiological research, the details are beautifully clear. Take the protein channels that are modelled in the heart pacemaker model described in Chapter 5. We now understand in immense detail what happens in such proteins. We can pinpoint where exactly the atoms sit that carry electric current through them.

This is what you get with the sort of top-down approach that physiology adopts. Our understanding of muscle contraction started with the anatomical identification of the muscles of the body. Then we managed to demonstrate that the muscles do not work by nerves injecting or removing fluids from them. They work instead by nerves secreting a chemical that excites electrical activity in the muscle. This in turn stimulates the production of calcium signals inside the muscle cells. The calcium signals excite certain proteins. These contractile proteins, as they are called, slide over each other using a kind of ratchet mechanism at the molecular level. We understand the finest details of this ratchet mechanism. It was Holmes' report of those details that so impressed Wolpert and me at the Novartis meeting.

This is the measure of reductionist biology's success over the last few decades. But let us now return to Lewis Wolpert's question: 'How much more do you want to know?' Holmes answered this at the molecular level by saying that he thought there were more molecular structures to be determined. I would want to add that we also want to know how it all works at higher, integrative levels. To demonstrate this, I will switch back to my own field of work: the heart.

The virtual heart

Heart muscle cells work somewhat differently from skeletal muscle cells, but the two also have a lot in common. The molecular processes involved in movement are essentially the same: two types of contractile protein slide over each other. The differences lie primarily in the way in which the cells are controlled, and how they operate in the organ as a whole. In the heart, all the muscle cells are connected together in a highly ordered and intricate way. So it absolutely is not possible to understand the workings of the heart as an organ at a purely molecular level. All the muscle cells interact. How they do so is crucial. This is what determines whether blood is pumped to the rest of the body, and therefore whether we live or die.

As with other organs and systems, the intricate details of how the heart is organised must be laid down during embryonic development. Ideally, therefore, a full quantitative understanding of the organ cannot be achieved without detailed knowledge of its development. This would provide us with clues to the logic behind its evolutionary success. Although much is understood about the development of the heart, we cannot yet model this process on the scale we would wish. This is partly because we don't know enough. Also, even if we did, such a project would be beyond what is possible mathematically (see the story of *Blue Gene* earlier in this chapter).

So here again we need to take a 'middle-out' approach. The 'middle' in this case is the organ itself.

I was privileged to see the beginnings of this approach about 15 years ago while spending time as a Visiting Professor at the University of Auckland in New Zealand. Peter Hunter in Engineering and Bruce Smaill in Physiology were collaborating on a painstaking project. Millimetre by millimetre they were recording the location and orientation of the muscle fibres in a dog heart. After years of work, they had millions of data points that could be arranged into a grid structure mimicking that of the original heart. The top parts of Fig. 6 show a more recent version of this approach applied to a pig heart.

Such a massive collection of data is important in its own right. It enables us to create virtual anatomical models that we can use, for example as a teaching tool. But the Auckland team had far larger visions than this. They assembled the database in the form of computer software that was also capable of incorporating the contractile (mechanical) behaviour of the muscle fibres, the circulation of the blood in the heart (middle row of Fig. 6) and the electrical properties using cell models of the kind described in Chapter 5. Future

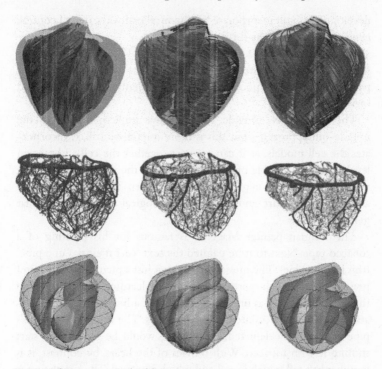

FIG 6. Three stages in building a computer reconstruction of the heart (the 'virtual heart').

TOP: modelling the orientations and locations of muscle fibres in the ventricles of the heart. (From Stevens and Hunter 2003, reproduced with permission.)

MIDDLE: modelling of the blood vessels of the heart at three points during a single heart beat:

        left: just before contraction, when the heart is most relaxed;

        middle: during contraction, when the heart is most active before ejecting the blood;

        right: at the end of ejection of blood, before relaxation begins.

        (From Smith *et al.* 2001, reproduced with permission.)

BOTTOM: three stages in the electrical excitation of the heart. (From Tomlinson *et al.* 2002, reproduced with permission.)

developments will incorporate biochemical pathways, neural control, gene expression . . . the list is a long one.

The result is the first virtual organ, a virtual heart. This project is now being pursued internationally. A number of teams in different parts of the world are pooling their data, mathematical modelling, and ideas.

The cell models provide an impressive implementation of the middle-out approach—but this is more impressive still. It incorporates the cell models, so it can connect together the cell and protein levels. It is also capable of reaching down to the genetic level. It can represent differences in gene expression patterns as between a normal and a diseased heart, and also the physiological effects of particular genetic changes.

The Belgian painter Magritte is famous for his painting of a tobacco pipe. Next to it he painted the text 'ceci n'est pas une pipe' (this is not a pipe). The message was: this is just a picture. Well, when I present movies of the virtual heart during lectures I sometimes add the text 'ceci n'est pas un coeur' (this is not a heart). But my motivation is a bit more pointed than Magritte's. The thing is, Magritte's pipe is accurately depicted, but no-one would be inclined to start stuffing it with tobacco. With movies of the heart, by contrast, it is very hard to tell which is real and which is virtual. The simulation is that convincing.

But how complete is it? Is the impression of 'reality' correct?

We don't yet know exactly how many genes are expressed in the heart, but the number is unlikely to be less than 5000. Even the most complicated of the heart cell models include no more than about 100 protein mechanisms. At a rough guess, therefore, we are representing only around 2% of the genes involved. Nevertheless this generates an immensely convincing reconstruction of the electrical and mechanical behaviour of the heart. This is a good indication of how modular nature can be. But it is also a reminder of how much further such a project has to go. And then we can start on similar analyses of the other organs and systems of the body.

Breaking Humpty-Dumpty down into his smallest fragments, the genes and proteins, was difficult enough. But it was probably the easier half, the reductionist half, of the challenge facing biological science. Putting him back together again is going to be more difficult still. This is part of the exciting challenge that faces systems biology.

7 § Modes and Keys: Cellular Harmony

It is not so obviously false as is sometimes made out.

John Maynard Smith (1998) on 'Lamarckism'

The Silmans find some tropical islands

Charles Darwin was a man for islands. What he liked about them was their isolation. He did not enjoy isolation for its own sake, of course, but he did appreciate the chance to study species that, being isolated, had developed in unique ways. This is how he made some of the key discoveries that led up to his theory of evolution by natural selection. He was studying forms of life, particularly birds and tortoises, in the Galapagos. The various islands in this archipelago off the coast of South America were more or less cut off from one another. The animal species on these islands had consequently diverged from each other.

It took time for the reasons for this divergence to become clear. During the voyage itself, Darwin was so unaware of their immense significance that he didn't even record the islands on which his specimens were obtained. It wasn't until he got home that the ornithologist and illustrator, Gould, identified his finches for him. Only then did he begin to appreciate the variation on the islands.

With that example in mind, I should like to offer you another story of the Silmans. This time, they discover some islands on earth where very strange things are happening. As in Darwin's case, it takes time for them to appreciate the significance of what they find.

We have said that Silmans are silicon-based equivalents of humans, with a similar degree of intelligence. So it would be most natural to imagine them as being of similar size to us. In fact, however, their silicon nature enables them to exist in different sizes, some of which are so small that they could even become parasites of animals on earth. For these Silman species many earth organisms are so enormous to their eyes that they hardly recognise them as organisms. Their investigation of earth biology is therefore stuck at the level of cells.

Like Darwin, they discover an archipelago of fairly similar tropical islands. The temperature all year round is a sweltering 37°C. Let's assume, too, that, like their larger distant cousins in Chapter 3, and despite their tiny size, they have the technology for sequencing DNA. On the first of these islands they discover a rich diversity of species. There are viruses, bacteria, and eukaryotes (species made of the same kind of cells as yours and mine: cells with nuclei and chromosomes, mitochondria, and ribosomes). So, they get to work with their DNA sequencing machines.

They discover a wide range of different genomes in the viruses and bacteria, but they also find about 200 species of eukaryotes in which the DNA genome sequences are all exactly the same! Yet, the species are quite clearly different. Some move and change their shape. Others are ciliates with waving hair-like cilia in various patterns on their surfaces. Yet others send out long lines of communication. They each live in colonies of rather similar characteristics.

The strange thing about these species is that when they reproduce (which they do asexually) they pass on their acquired characteristics, those of the colony in which they live, to the daughter cells. Even though they have exactly the same DNA, the same set of genes as all the other 199 species on the island, they succeed regularly in transmitting their acquired ability to move, secrete, transmit, and so on.

One of the Silmans knows something about the history of biological thought on earth. So he immediately appreciates that they have made a discovery of dramatic significance.

'You know,' he explains, 'there was a big argument in the nineteenth and twentieth centuries here on earth. Some said that the inheritance of acquired characteristics was impossible, and others (including, curiously, both Darwin and Lamarck, to differing degrees) thought it was possible.'

'Later', he continues, 'they resolved this argument in favour of saying it was impossible because the genome DNA code was never found to be altered by such acquisition by individuals. Whatever happens to each organism in its interaction with the environment is not passed on to its offspring in the DNA code. What they called neo-Darwinism was based on this principle, which they called the central dogma of biology. So what we have found would greatly surprise them. We must investigate the mechanisms involved.' Another Silman recalls the role of the Galapagos in Darwin's work and proposes that they extend their enquiry to another island so that they can make a comparison. What they then find is even more astonishing.

The second island also presents a rich diversity. Here again there are viruses and bacteria with different genomes. Just as on the first island, there are around 200 species that all have the same DNA in their genomes. But this DNA is different from that found in the equivalent group of species on the first island! They rush to the next island. Same story again. Once more, around 200 species of eukaryotes carry identical DNA, but it is different from that on the other islands.

One of the Silmans knows that mitochondria also have DNA that is inherited. So they do some research on mitochondrial DNA sequences. Again they find the pattern differs from island to island but is always the same on a given island.

They are accomplished molecular biologists so they eventually work out what is happening. On each island, each species is imprinting a chemical pattern on the DNA, even though the DNA code itself is standard across the island. As a result, when the cellular machinery expresses the genes the same pattern of expression occurs in the daughter cells as in the parent cell. Even though the DNA sequences are identical for all species, the expression patterns can be completely different. It is these that produce the differences between species. This inheritance of expression patterns occurs throughout the generations. It is completely robust.

Some of my readers may be surprised to know that this is in fact chemically possible, as I will explain a little later. Despite appearances, the Silmans' discovery would not in fact surprise molecular biologists and geneticists today. But they do not know this. They think they have discovered something new and different. So—what to do now?

The Silman who first suggested investigating different islands has another bright idea. 'You know,' he says, 'what we have found fully explains the transmission of acquired characteristics within each species on an island. But every island has a similar set of species. Yet the species of a certain type on one island has different DNA from the equivalent species on any other island. Why is that? How did those differences evolve? That must be the key to understanding what is going on here.'

Since these Silmans are tiny, they also appreciate time rather differently from us. Our year is like 100 years to them. It therefore takes them considerable time to work out the next clue in the mystery. Taking careful measurements, they notice that the islands are not completely immobile. Very, very slowly they move with respect to each other. This doesn't surprise them. They know about tectonic plates and how even huge continents on earth have shifted around.

After what for them amounts to several hundred years, they observe two of the islands coming very close together. So close that a bridge can be projected across from one island to another.

First, there are some problems with connecting the bridge. It seems to require many fumbling attempts before a mooring occurs to a tunnel on the other island, and even then the bridge slides backwards and forwards as though it is still not sure that it is securely in place.

Then something perfectly mind-blowing happens. Suddenly, without warning, there are simultaneous earthquakes on the connected islands. The vibrations and aftershocks almost throw them off the islands. There are also blood-curdling shrieks that, had they come from a Silman, would surely signify great pain.

They then witness the most remarkable event of all. It concerns one of the 200 species, which has until this point appeared to them rather dull. Members of this species didn't seem to do anything (that is, didn't seem to have a very specific phenotype). Now, suddenly, millions of them set off from one island and start swimming rapidly in a fast-flowing stream across the bridge, which is a kind of enclosed aqueduct, to invade the tunnel.

The earthquakes cease, calm returns, while the invaders race deeper into the tunnel. Most die in the race. Only one of them survives to attach itself to a large cell on the other side. At this point, an electric shock-wave transmits across the cell surface, and all the DNA of the motile species suddenly passes into the large cell. Many years later, the island divides into two and another, smaller, island appears. It also has many different species with yet another identical genome.

The Silmans' mistake

You understand, of course. The millions of tiny motile cells crossing the aqueduct are sperm cells. One of them encounters an egg cell and fertilises it. The rest of the 200 species from each island are the differentiated cells of a human body, which of course also hosts viruses and bacteria. The Silmans' mistake arose because of their size. Being so small, they mistook each human body for an island, and all its various cell types, all 200 of them, as individual species.

But their mistake has a large element of truth. To the cells themselves, the human body is rather like an island on which, from an evolutionary point of view, they are trapped. Organisms not only capture and enslave genes, viruses, and bacteria, they also capture whole cells.

Consider the story with this in mind. It is all standard biology, except for the description of differentiated adult cells as different species. So the Silmans' discoveries would not surprise any scientist on earth. Yet, relating the story in this form seems shocking. Why? Largely because 'Lamarckism', the inheritance of acquired characteristics, is widely and wrongly thought to be the antithesis of Darwinism. I will clarify this matter at the end of this chapter.

Genetic basis of cell differentiation

At the level of the organism inheritance of acquired characteristics between generations seems to be very rare. But in multicellular organisms there is a level at which a form of such inheritance is rampant. All cells in the same body have the same complement of genes,[1] inherited from the original gene combination formed from the sperm–egg fusion. Yet they are remarkably different from each other. Bone cells differ from nerve cells, pancreatic cells from skin cells, liver cells from heart cells.

How are these inheritable differences explained? It all comes down to the fact that, although the DNA code is the same in every cell, different genes come into play in each type of cell and genes are also expressed differently—which genes get expressed when is different, and how they are expressed is also different. The musician always plays the same 30 000-pipe organ (the genome), but he plays it quite differently in each cell type.

And these specialised cell types in the body have a remarkable feature. When they divide to generate new cells, they transfer the information concerning their acquired pattern of gene expression to the daughter cells. This is called epigenetic inheritance. It does not depend on differences in DNA sequences.

The DNA is of course passed on to the daughter cells, but it is not this that makes them liver cells, skin cells, or whatever. Rather, it is an additional level of patterning that is imposed on top of the DNA itself that makes the difference here. The DNA carries a form of chemical marking which is different for each type of cell. This ensures that the patterns of gene expression specific to each cell type are transmitted down the generations of cells.

One of these marking mechanisms involves a chemical process called methylation of cytosines (the C of the DNA code). This particular chemical mechanism of gene imprinting is understood. There are others that are not yet well understood. In addition to gene imprinting, cells also receive signals from the body itself. A group of

[1] The exception is the germ cells, which have only half the DNA code.

pancreatic cells transferred into the brain, for example, will continue to act as pancreatic cells, but outside the body they will lose pancreatic function. The gene imprinting, however, remains. This seems to be permanent.

The processes involved here are fascinating. Embryonic skin cells that happen to be in the region of the body surface where an eye will be formed are induced to change their expression imprinting so that they turn into the sort of cells that form a lens. The conductor can, as it were, transform a lowly flute player into a magnificent trumpeter!

It did not have to be like this. About a century ago, one of the originators of modern ideas about genes, August Weissman, thought it more likely that, instead of differentiated cells having the full complement of genes, they would only have those needed to express the function of the cell type. So a liver cell would have a different complement of DNA in its nucleus than a heart cell. This idea was an older version of 'genes for everything'. There were genes for heart cells, genes for pancreatic cells, genes for nerve cells, and so on.

At the time this seemed the simpler, more likely, idea. If you assume that each function must correspond to a set of genes, then it would make sense to separate out the genes according to the functions they code for and to include in cells with a certain set of functions just the genes that those functions require. That solves the problem of how the cell's type-specific characteristics are transmitted. We now know that things are different. And we know why—the relationship between genes and function is complex. Genes play roles in many different functions. It is patterns, not just individual genes, which matter.

What distinguishes a heart cell from, say, a pancreatic cell? It is not so much which genes are activated. It depends on the extent to which they are activated, compared to other genes. Nowadays the existence of these so-called epigenetic inheritance mechanisms in differentiated adult cell types is well-established. So much so that I suggest we need to reverse the usual thinking on this question. We should not be asking how it can happen in adult differentiated cells, but rather how such a natural and useful mechanism comes to be suppressed in the germ cells. Perhaps it is not completely suppressed. Conceivably, such chemical marking of the genes is involved in the rare forms of 'Lamarckian' species inheritance discussed in Chapter 4.

The interesting question, as Maynard Smith fully realised, is why the phenomenon is so rare in germ-line inheritance. There is no obvious reason why it could not have been very common. Cells that are trapped in the same multicellular organism use it freely. But, for the central biological dogma to be correct, inheritance via germ cells must not use this mechanism.

But that is to put the cart before the horse. Why did evolution make the central dogma almost invariably correct? Why does a 'Lamarckian' form of inheritance flourish within the colonies of *all* the cell types in multicellular organisms with the *sole exception* of their germ cells?

Maynard Smith (1998) has offered one possible explanation. He writes: 'most phenotypic changes (except learnt ones) are not adaptive: they are the result of injury, disease and old age. A hereditary mechanism that enabled a parent to transmit such changes to its offspring would not be favoured by natural selection.'

I am not so sure about this. Most genotypic changes (mutations) also are not adaptive. They are usually quite as bad as injury, disease, and old age, which is why natural selection rejects most of them. If natural selection can be so effective in filtering out bad genotypic changes, it could have been effective for bad phenotypic changes. 'Lamarckian' inheritance would not exclude Darwinian selection. It would complement it, providing yet another source of diversity.

If nature could have used this mechanism, it would surely have done so. So we return to the question: why is it so completely suppressed in the germ cells? My hunch is that the answer may lie somewhere in the development of biological complexity above the cellular level. It seems to be when cells come together in the large colonies that form multicellular organisms that epigenetic inheritance is given free rein.

Modes and keys

For the great majority of evolutionary history, multicellular species did not exist. Almost certainly the first two billion years did not see them emerge, and they only appear in great profusion at the time of the Cambrian explosion around 530 million years ago. So they have been around for only 13%, perhaps at most 15%, of the period of four billion years that life is thought to have existed on earth. We can't be totally certain of the timing since the very earliest multicellular forms would have left no fossils. But it was still only relatively recently that multicellular forms emerged.

Their characteristic is what I will call cellular harmony: in a healthy organism they must co-operate in a harmonious way in the interests of the whole, despite the fact that they also have their own 'selfish' interests which, if given free rein, lead to diseases like cancer. I say 'must co-operate' since, like the genes (see the quotation at the beginning of Chapter 1) 'they are all in the same boat'. However the co-operation has arisen, it is necessary.

There is an analogy with the development of musical modes and keys. For the great majority of human musical history, rhythm and melody were the key features.

What we call the medieval modes were perfectly adequate for music involving a single line of melody with rhythmic back-up. These were used for Gregorian plain chant in churches, and almost certainly in popular non-religious music. For example we can iden-tify the use of these modes in the beautiful and erotic songs of the eleventh- and twelfth-century troubadours. The same is true for many early forms of music around the world, although the modes (the arrangement of notes within an octave) may differ significantly from Western medieval modes. There were many possible variations in the demands of this kind of music. Traditional Japanese, Korean, or Indian music sounds very strange to Western ears.

But then the fashion changed. There was a demand for several voices or instruments singing and playing, not in unison, but in all the complex forms of interrelationship that we call polyphonic harmony. It was in response to these demands that modern musical keys developed. The great advantage of these keys is that, provided that each musical part is written in the same key, and certain rules concerning intervals are followed, the results are very harmonious. The medieval modes, however, didn't disappear. They became assimilated, just as unicellular organisms did not disappear when multicellular forms evolved.

Multicellular harmony

Can we get a clue from this metaphor? Could epigenetic inheritance and its exclusion from the germ cell line be a requirement of multicellular harmony? Must all the cellular music be in the same key?

The exact number of cell types in a human is debatable. It is all a question of definition. A project that seeks to model all the cell types in the body (the Human Physiome Project) estimates that there are around 200, all with completely different gene expression patterns. There would be even more if one took account of finer variations, such as those that occur in various regions of the heart and which are thought to protect the heart against fatal arrhythmias.

So we have a choir, or an orchestra, of around 200 individuals. The precise number is not too important. The important fact is that it is large and that the range of patterns of gene expression is therefore also large and varied. Their patterns must also be harmonious in the context of the organism as a whole. They are all in the same boat: they sink or swim together. Disturbing their harmony would have serious consequences. It was arrived at after more than two billion years of experimentation. It may not be perfect, but most of the time it works.

We have also seen that each cell type is so complex that the great majority of genes are expressed in many cell types. So it makes sense that all the cells in the body have the same gene complement, and that the coding for cell type is transmitted by gene marking, rather than by gene complement.

This may be a helpful context in which to approach the question of germ-line inheritance. Consider what would happen if germ-line inheritance reflected adaptive changes in individual cell types. Given that all cell types derive ultimately from the fused germ-line cells, what would the effect be? Clearly, it would be to alter the patterns of expression in nearly all the cell types. There would be no way to transmit an improvement in, say, heart function to the next generation via gene marking of the germ cells without *also* influencing the gene expression patterns in many other types of cell in the body. And of course there is no guarantee that what is beneficial for a heart cell will be so in, say, a bone cell or a liver cell. On the contrary, the chances are that an adaptation beneficial in one cell type would be deleterious in another.

We encounter this problem frequently in the pharmaceutical industry. We develop drugs for an action on one type of cell in the body, such as a kidney cell. Unfortunately these drugs also influence many other cells. Cells that express the same genes cannot avoid also being sensitive to the same drugs. The result can be a serious disturbance of the delicate harmony between the cells of the body. In the case of a drug we call this a side-effect.

Side-effects present a serious problem, indeed they are often debilitating and sometimes fatal. The same would apply to acquired-characteristic genetic changes. Much better, therefore, to let the genetic influences of natural selection be exerted on undifferentiated cells, leaving the process of differentiation to deal with the fine-tuning required to code for the pattern of gene expression appropriate to each type of cell. Natural selection of the germ-line code determines the overall harmonic key, while epigenetic inheritance of individual cell types determines the parts they play.

If this explanation is correct, we would not necessarily expect it to be 100% effective. It is conceivable that some germ-line changes in gene expression patterns might be so beneficial for the organism as a whole, despite deleterious effects on a few cell lines, that the result would favour selection. This could explain the few cases where germ-line 'Lamarckian' inheritance seems to have occurred (Chapter 4). It also motivates the search for other cases. The prediction would be that it will occur in multicellular species only when beneficial to overall intercellular harmony. It might be more likely to occur in simpler species. That makes sense in terms of the few examples that we have so far found (Maynard Smith 1998).

A historical note on 'Lamarckism'

The story in this chapter may appear shocking for reasons that are somewhat different from other chapters. Why is this the case?

The answer is a series of hang-ups and serious misunderstandings about Lamarck and 'Lamarckism'. The word itself is almost banned from current biological thought and, when it is used, is almost always a term of denigration. To be accused of 'Lamarckism' in biological science is almost as bad as creating a Maxwell's Demon (breaking the laws of thermodynamics) in the physical sciences.

There is a widespread perception that Darwin and Lamarck fought over this issue of the mechanisms of inheritance. The truth is that neither knew anything about such mechanisms. Darwin's great achievement was to propose a process by which, whatever the mechanisms of inheritance, blind selection could generate new species. He rejected Lamarck's idea that evolution has an inherent drive. Both Darwin and Lamarck absorbed the idea of inheritance of acquired characteristics (use and disuse) from others, since it was commonly assumed from time immemorial.

It is true that Darwin was less enthusiastic about it, but it is an unfortunate accident of history that Lamarck's name has been universally identified with an idea that he did not invent, and that this idea is now seen as the antithesis of Darwinian evolution by natural selection. Neither Darwin nor Lamarck would recognise this travesty of the history of biological thought, although for quite different reasons, Darwin was extremely dismissive of Lamarck when he described Lamarck's 1809 book (Lamarck 1994) as 'veritable rubbish'.

We are now completely saddled with the term 'Lamarckism' for the inheritance of acquired characteristics. I will therefore use it here, having fully acknowledged the injustice done to Lamarck himself, who would be better and more correctly remembered for having first introduced the term 'biology' and established it as a separate science.

Mayr's monumental book (1982) redresses the balance of this particular history. For French-speaking readers, so does Pichot (1999). This history is the reason why in this book I always use 'Lamarckism' in quotes. The muddle is such that it is quite difficult to define precisely what is meant. I follow Maynard Smith (1998: 11) in using the term to refer to mechanisms of inheritance that contradict the strict assumptions of neo-Darwinism. Maynard Smith puts all the examples I refer to in this category, including inheritance of cell differentiation.

This is also a suitable point at which to refer to the important contributions that have been made by modern French scientists on many of the issues in this book. In addition to Pichot, the book *Ni Dieu ni gene* ('Neither God nor genes') by Kupiec and Sonigo (2000) was influential. And I have already acknowledged the debt to François Jacob's *La Logique du vivant*, although I clearly dissent from his notion of a genetic program.

8 § The Composer: Evolution

> We don't have a theory of interactions and until we do we
> cannot have a theory of development or a theory of evolution.
>
> Dover 2000

The Chinese writing system

There are 20–30 000 genes in the human genome, which combine to produce effects at all levels of the organism. The number of ways they can combine is enormous, as we found in Chapter 2. There I used the comparison with a pipe organ.

There is another human invention that has around the same number of components. This is the Chinese writing system, used also in Taiwan, Japan, and Korea, and formerly in some other East Asian countries. Chinese characters are stylised pictures that fit into the space of a square. Strings of these squares can be used to represent sequences of words in a language. Each picture has a meaning; in some cases the meaning is easily related to the picture.

The character for a mountain, for example, is a series of three peaks, the largest in the middle. We can, with a little imagination, see in it something like a mountain range. Historically it was quite like a child's drawing of mountains. Here is its modern version in two different fonts, square and cursive.

山　山

In most cases, however, the meaning is less easy to see. It requires considerable knowledge of the conventions. So people who have not grown up in a country that uses the characters have great difficulty learning them. The initial impression is of an arbitrary jumble. Westerners naturally wonder how people can be so crazy as to use this as a writing system instead of an alphabet.

The wonderment increases. To learn the roughly 2000 characters necessary for basic fluency is hard enough. Then one discovers that historically there have been around 40 000 of them,[1] of which nearly 10 000 survive in modern educated usage. Suitable for an amateur, my Chinese, Korean, and Japanese dictionaries are more modest. Each has only around 5000–7000 characters.

Some are fiendishly complex. The character used to indicate 'imperial', and which in origin is a fabulous mythical bird, requires 30 separate strokes of a brush or pen to reproduce it.[2] Here it is using the same two fonts as for 'mountain'.

The number of possible ways in which 30 strokes could be arranged to produce characters is astronomically large; 30–40 000 is actually a very tiny number compared to what would theoretically be possible.[3]

This already suggests that there is much more order in the system than may appear at first sight. And so there is. The character reproduced above is formed of just four simpler characters, one of which is repeated, so there are really only three. They have simple meanings: thread 糸, speak 言, and bird 鳥.

[1]   The Kangxi zidian (康熙字典), a 42-volume dictionary published in China in 1716, lists over 40 000 characters.

[2]   This character is nearly obsolete and is now encountered most commonly in the name of a famous Buddhist monk, Shinran 親鸞 (1173–1262).

[3]   If each pen or brush stroke can have five different orientations (vertical, horizontal, two diagonals, and no stroke at all) then there are about 10^{21} (1000 billion billion) possible characters with 30 strokes, even without taking account of stroke length. If one adds, say, four different stroke lengths, the number rises to 10^{37}. Since the actual number is only of the order of 4×10^3, the ratio of possible to actual is around 10^{33}.

All the 30–40 000 known characters are formed from combinations of these basic elements. There are only 200–300 of them; most are characters with specific meanings in their own right. An even smaller number, around 100, appear very frequently. 'Thread' and 'speak', for example, appear in hundreds of characters.

So the first stage is to learn a hundred or two of the simpler characters. Thus equipped, one can already see calligraphic patterns, sometimes reflected also in meaning, in all the rest. This is what we would call a modular system. When first encountered, the complexity seems overwhelming. Then one finds that simpler elements are re-used over and over again to create that impression.

Modularity in genes

By now, the reader must be familiar with the games played in this book, and I hope you have enjoyed them. If you guessed that this digression into modularity in Chinese characters is a prelude to exploring the modularity of life and how it has evolved, you will have guessed right. As with Chinese characters, so with life: here again is a modular system. And its modularity is the key to understanding how it has evolved.

Genes, we know, are long stretches of DNA code. Each is built up of smaller modules, like a mosaic. We don't know exactly how many such modules there are, but it looks as though there may be as few as a thousand or two. So these basic modules must be shared by a large number of genes.

Gene codes share another feature with Chinese characters. Although historically the modules may have had simple functions (meanings), the system as a whole is not at all simple and straightforward. Indeed, as we sit down to try and work it out, our first impression is of a frightful mess. The trouble is: evolution. As the genomes (languages) have evolved, so the functions (meanings) have changed. And they have changed in ways that are frequently arbitrary with respect to the original functions (meanings).

Genetic and cultural forms of evolution share this messiness or, to use a less derogatory term, inventiveness. For it is through a complicated series of bodges that nature has arrived at the huge diversity of life as we know it. Tangled intricacy is the mother of nature's invention.

The idea of metaphor is important here, too. A metaphor changes the applicability of a word or phrase. On that basis, we can say that, as the genome has developed, nature has switched from one metaphor to another. It has plundered the treasure chest of old DNA modules to form new combinations and to give old genes new functions. Species with ears, eyes, legs, wings generate such functions using genes that started out in creatures that never had any such ambitions. Nature rarely creates completely new modules; so many of the modules are extremely ancient. They evolved very early. The result is that the genomes even of species that are far apart in the evolutionary tree have a lot of sequences in common.

More than 99% of our genes have a related copy in the mouse. Despite 500-million-odd years of evolutionary separation, half the genes in a sea squirt correspond to ones we have, too.

We should not be misled by this. Popular media commentators parade the small percentage differences between human and mouse genomes as if to say that you and I are at bottom very little different from a mouse. It is a shame they have not looked at the calculations in Chapter 2. Tiny differences in sequence can encode enormous differences in function. All the differences in human races worldwide are thought to be coded by only 0.1 % of the human genome (International_HapMap_Consortium 2005): just a few million variations, of which maybe only half a million may be important in assessing health and disease tendencies. These figures seem tiny taken as a fraction of the whole. But they become huge as soon as we consider the number of ways in which the affected genes can interact between themselves and the rest of the genome. This is the basic reason for the relatively unsuccessful track record genetics has in dissecting complex disease traits.

Gene–protein networks

We sometimes say things like, 'This gene does such-and-such.' But that sort of statement tends to be misleading. A gene will do one thing in one set of circumstances and another if circumstances change. Indeed, it might be more helpful to avoid saying that genes do anything at all; it is more that genes are used. They operate under control. There is regulation, as biologists say. Conditions in the cellular environment will switch a gene on or off to varying degrees.

Genes are controlled by proteins. Those proteins in turn are coded for by other genes. Those other genes can in turn be switched by other proteins coded by yet more genes. The system depends on massive networks of such gene–protein–gene–protein . . . etc. . . . interactions. These are often called gene networks. Gene–protein networks would be better (recall the example in Chapter 5 of how circadian rhythms are generated).

The terminology matters. Talk of 'gene networks' adds to the impression of a program that is all in the genes and controls the development and maintenance of life. There is no such program (Chapter 4). The genes cannot do what they do without the proteins. And the proteins are not free agents, either. They respond to influences from across the rest of the organism and ultimately from the environment, too. This is how what we are calling 'downward causation' happens. So even if we speak of gene–protein networks, we must be careful. These are not networks that operate independently of higher-level processes.

Complex networks of this kind have properties that are important in both evolution and development. Their modularity is crucial. This allows them to be re-used in many different situations. It also means their switching can be changed without upsetting the network itself.

Consider a gene that is involved in switching on the network responsible for the development of a mammalian eye. Suppose we transfer it to an insect. What happens? Insects' eyes are structured very differently from mammals'. No matter: the transferred gene proceeds to switch on the development of the insect eye.

Again, consider the insect genes that switch modules involved in leg production. Suppose we transfer it to a different region of that insect's genome. Say we choose the region where one normally finds the genes that switch modules that trigger the production of a wing. What happens? A leg appears in the wrong place.

How do we understand such experimental results? Gene–protein networks constitute modular subsystems that are remarkably robust and adaptable. It is not an exaggeration to say that the discovery of such 'organising' networks has transformed our view of evolution. Before the genetic revolution, earlier anatomists and embryologists had pointed out the resemblances in form between animals as different as snakes, insects, crabs, and humans. They identified a certain segmental organisation that they saw was common across a range of species. Our understanding of the genetics of these phenomena largely confirms their ideas. The segmental organisation of the human backbone does indeed have the same origin as that of a snake. There is an underlying molecular network of control. It is based on what are called *hox* genes.

A *hox* gene may 'control' networks involving thousands of other genes and proteins. So it is often referred to as a 'master' gene. Here again we see how social and psychological assumptions intrude into our scientific work. The *hox* gene doesn't really 'know' what the network does, let alone impose its will on the network. Its role is critical in an important biological process, but it is not the master of that process; it is just the trigger. It triggers the operation of a large and complex network. But it does so 'blind'. If it is put in a position to trigger another network in another species with the same 'trigger' pattern, it will do so.

Fail-safe redundancy

The second important property of these networks is robustness, sometimes also referred to as redundancy. Redundancy is the necessary basis of robustness.

Suppose there are three biochemical pathways A, B, and C, by which a particular necessary molecule, such as a hormone, can be made in the body. And suppose the genes for A fail. What happens? The failure of the A genes will stimulate feedback. This feedback will affect what happens with the sets of genes for B and C. These alternate genes will be more extensively used. In the jargon, we have here a case of feedback regulation; the feedback up-regulates the expression levels of the two unaffected genes to compensate for the genes that got knocked out.

Clearly, in this case, we can even compensate for two such failures and still be functional. Only if all three mechanisms fail does the system as a whole fail. The more parallel compensatory mechanisms an organism has, the more robust (fail-safe) will be its functionality. Engineers, for example, use the same principles in building aircraft control systems.

Evolution has even more need for this kind of robustness than aircraft designers. It must be capable not only of designing a new 'aircraft' but of also doing so while the original one, and all the intermediates, continue to 'fly'.

Let's explore an example of this kind of back-up. Figure 7 uses the heart rhythm model we have already seen (in Chapter 5). As in Fig. 3, the top trace shows the cell voltage as the model cell 'beats' over a period of 12 seconds. The middle and bottom traces show the activities of two protein channels. One is a sodium channel (middle trace). The other (bottom) is the same mixed cation channel as before. At the start of the pacemaker activity, the sodium channel is about six times as active as the mixed channel. So its oscillation during each cycle is much larger.

After two seconds, the sodium channel was reduced in activity by 20%. Given its large role, one might expect a substantial reduction in rhythm. In fact the rhythm change is so small that it can't easily be detected on this graph. What has happened? As the sodium channel is reduced, the mixed cation channel takes over its role. In this experiment, it nearly doubled its activity. In this way, it completely replaced the activity lost by the sodium channel.

Next, let us see what happens if the sodium channel is reduced still further. Let us reduce it by 40% at 4 seconds, 60% at 6 seconds, and 80% at 8 seconds; and then, at 10 seconds, let us knock it out completely. What happens? There is now a detectable fall in frequency. But the effect is small. This is because the mixed cation channel now carries as much current as was formerly carried by the sodium channel.

For an important function like this heart rhythm there are several

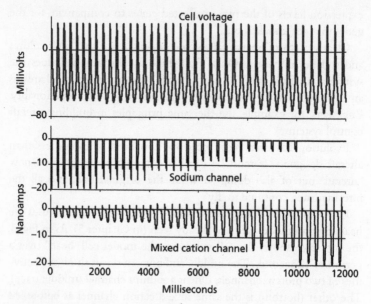

FIG 7. Simulation of a 'gene knock-out' effect (based on Noble *et al.* 1992) showing a mechanism of robustness in the heart's pacemaker. As the sodium channel protein is progressively reduced and eventually 'knocked out' (middle trace), the mixed cation channel protein takes over its role (bottom trace). The result is that the rhythm of the heart is maintained.

back-up mechanisms. There are so many, in fact, that physiologists have argued for decades about 'Which is the real pacemaker mechanism?' The answer is 'It depends'. It depends on species, it depends on conditions, it depends on control mechanisms.

Suppose you are running a race, or anticipating meeting a lover. Your heart beats faster. The balance of protein activity changes. Or suppose you experience a deep shock. The heart slows and you nearly have heart arrest. Again the balance changes, but in a different way. This rhythm generator is essential for life. If it starts to fail or cannot entrain the rest of the heart in a regular rhythm, sudden cardiac death looms.

It is natural that back-up systems should come into play.

Faustian pacts with the devil

It may not be easy for one species to change into another. Imagine fish evolving to become dry land species. It is easy to see how problems arise as fins become legs and lungs are needed. Somehow, the evolutionary process got there. How? Almost certainly, the features of modularity and redundancy were critical. It is possible to slot pre-existing gene–protein networks into new control networks without upsetting them too much. That is modularity.

Suppose mutations occur in some mechanisms, and eventually they are selected for functions quite different from those they originally supported. How, then, to maintain the function that they originally served? What were back-up mechanisms now become primary. That is the value of redundancy. This is the basic explanation for how nature can modify its 'aircraft design' while still ensuring that the aircraft continues to fly.

Of course, evolution is a blind process. Looking back, we can say that particular modules and redundant mechanisms were critically important in particular evolutionary developments. That is not to say it was all worked out in advance. Just as the heart oscillates without a specific oscillator to drive it, so evolution works without a master plan. The process will sometimes, in fact frequently, lead down blind alleys (that way lies extinction—the fate of the vast majority of species) or into what I call 'pacts with the devil'.

These are what we, with hindsight, would call design faults, but which, from an evolutionary point of view, are the inevitable price paid for many successful developments. They resemble Faust's pact with the devil. Recall that, in this story, Faust secured years of unlimited knowledge and power from the devil, but at the price of giving the devil his soul. The key to this kind of pact is that it is eventually fatal but for a long time it is of great benefit. This is just the kind of pact that nature stumbles upon when it finds a good combination of genes to generate a function, at a price that may eventually be fatal. Evolution may take little notice of the fatality in individuals, particularly if it occurs well after the reproductive period of life.

One of the causes of fatal heart disease is the consequence of such a pact. We saw the beneficial consequences of this pact in Chapter 5, where we discussed the potassium channel protein whose function ensures a low energy requirement for electric excitation of the heart. That was the good news; now for the bad news. By making the heart energy efficient, it also made it fragile as an electrical mechanism.

There are potassium channels that switch on during each heartbeat in such a way as to ensure that the heart returns to its original state after the beat. Unfortunately, the proteins to which evolution has entrusted this crucially important task are amongst the most promiscuously reactive in the body. For example, they interact with around 40% of the chemical compounds that the pharmaceutical industry comes up with in its search for new therapeutic drugs. These drugs therefore upset the electrical recovery of the heart and can kill.

The potassium channel proteins are susceptible in other ways, too. Quite a number of mutations in these channels and in other channel proteins can also create the conditions in which they cannot do their job. These genetic mutations predispose people to sudden heart death. They may live, unsuspecting, for forty, fifty, or more years and then suddenly, perhaps in the shower after an exhilarating game, they just drop dead.

Some Faustian pact!

Could evolution have avoided this situation? We don't really know. It might have hit upon potassium channel proteins that are not so promiscuous. It might have found other fine-tuning mechanisms (as it did for pacemaker rhythm) that make the system more fail-safe. But, most likely, it neither cared nor knew about these problems. If they affect only a small fraction of the population, and kick in after the main reproductive period, then there is no reason why selection should have filtered them out. And evolution certainly did not anticipate the arrival of the drug industry! Indeed, it could not anticipate anything. Remember that the process has no direction, and no foresight. The expressions 'caring about' and 'knowing about' simply refer metaphorically to what is likely to play a role in the process of natural selection.

The logic of life

The original (Greek) meaning of 'physiology' is 'life-logic' (physio-logos). In the Chinese characters used in Japanese, Korean, and Chinese, this meaning is even more evident. They are

生理学

which in sequence is 'life-logic-study'. So, does life have a logic?

Some evolutionary geneticists have argued that it cannot have. The evolutionary process is blind, imperfect, and subject to chance. Accordingly, life cannot have a design, or be perfect, or follow a rigorous logic.

In fact we can invert Paley's argument for the existence of God. He argued that if we found a watch in the desert, we could at least conclude that there was a watchmaker. So, finding life with all its intricate beauty and adaptation to its environment, he said, we must surely assume an intelligent creator. Now, by contrast, we can see that life is full of design faults, false trails, and imperfect compromises. We can still wonder at the intricate beauty of life on earth, but we no longer think that its logic is the best there could be.

Moreover, as we saw in Chapter 2, there might have been billions of other solutions to the problem of evolving living systems. If and when we do find life elsewhere in the universe, the chances of it being like us may be close to zero. Remember: there would not be enough material in the whole universe for evolution to have experimented with all possible combinations.

But none of these arguments mean that there is no logic at all. There may be billions of possible logics for life, and still it may make perfect sense to refer to what we see on earth today as having a logic of some particular kind—earth's own life-logic. It would be just one of those billions of possibilities; evolution might have taken all sorts of different turns at any number of the many corners it encountered during the billions of years; and still what we have got makes its own kind of sense.

The grand composer

That brings me back to the title of this chapter: evolution is the grand composer. It has orchestrated the music of the genes, the harmony of the cells, the symphony of the various stages of life. It has achieved this by serendipitously winnowing down the possibilities. In this way, it has continually adapted organisms to their environment, which of course includes other species. This logic, even if neither perfect nor designed, is there to be worked out. To achieve this, it will be necessary to reconnect physiology and developmental biology to theories of evolution. Patrick Bateson (2004) expressed this need, and that of moving away from the gene-centred view of evolution, when he wrote 'The decoupling of development from evolutionary biology and the attribution of intentions to genes could not hold sway for ever. Whole organisms survive and reproduce differentially and the winners drag their genotypes with them. This is the engine of Darwinian evolution and the reason why it is so important to understand how whole organisms behave and develop.'

This must be the ultimate goal for systems biology. At present we are only at the very beginning of the attempt to do this. We are looking towards a mature theory of biological systems-level interactions, but at the moment we have only a small glimmering of how it might be possible to develop such an understanding.

The task of systems biology is first to unravel these interactions and then to develop theories to account for them, and so to lay bare their logical underpinnings. Success in this endeavour is an essential prerequisite for developing a theory of development and a theory of evolution. We won't find that in genomes alone. Remember Dover: 'There are no genes for interactions as such.' It all has to emerge without there being a driver. The grand composer was even more blind than Beethoven was deaf!

9 § The Opera Theatre: the Brain

I think the secret of consciousness lies in the claustrum—
don't you?[1]

Francis Crick, 2004

We are approaching the end of the story. We have come a long way
from the first steps at the level of the genes, through to the grand
composer of the music of life, evolution. At each stage we discovered
that there is no single controller. The orchestra of life works without a
conductor. We might find that discovery relatively easy to accept for
viruses, bacteria, plants, and the simpler animals. But can this really be
true for the so-called higher animals, including man? After all, we
have a huge brain, with billions of nerve cells. It may well be the most
complicated thing in the universe.

Some readers will therefore already have concluded that there is an
obvious answer to the question 'What controls the processes of the
body?' Yes, the nervous system is certainly a central integrator and
controller of some kind. The question is what kind. Must we go along
with Crick, and many other biologists, in looking for a place in
the brain where it all, as it were, comes together in a central

[1] The full quotation is 'I think the secret of consciousness lies in the claustrum—don't you? Why else would this one tiny structure be connected to so
many areas in the brain?' (Francis Crick, 2004, quoted by V.S. Ramachanran, in
'The Astonishing Francis Crick', *Edge* **147** (18 October 2004, www.edge.org)).
The claustrum is a thin layer of nerve cells in the brain. It is very small and it
has many connections to other parts of the brain, but the details are of no
importance to the argument.

consciousness? Could a bit of the brain, like the claustrum, or any other part, do this?

And, if so, how does this conscious centre see what it sees, hear what it hears, feel what it feels? Does the nervous system serve up our sensations to it in a special form, converting the light, sound, and pressure waves into special qualitative phenomena (some philosophers and scientists call them sense data or qualia[2]) that exist inside our heads? This is an area where biology and philosophy strongly interact and, some would say, overlap. So how do biologists and philosophers think that we perceive the world?

How do we see the world?

There is a whole set of philosophical puzzles about the self, the brain, and perception of the world. One argument lies at the core of them all. We encounter it in various guises. Here is a story about it. The protagonists in the story are called ME and YOU. All you need to know about them is that they have the same mother.

While I am writing this chapter, I look closely at the print. I note that it is black and the page is white. So, I tell you that I am writing this with a black font on a white background. You reply that, yes, you understand what I have told you, but you nevertheless have a nagging doubt.

'How', you ask, 'do you know that I see black, or white, as you do? Perhaps, what I see when I read your book is what you would see as blue on pink, or green on ultramarine, or any one of millions of possible combinations, including colours that perhaps you have never seen! All we can really say here is that my perceived colours must relate to yours in such a way that we always agree on what to call what we see. We use the same names. But we may see differently.'

At first I respond naïvely. I say, 'Don't be silly. We both learnt at

---

[2] This term was originally introduced by twentieth-century philosophers to refer to the 'qualitative character of experience'. The singular is 'quale'. Earlier terms for this include 'sense impressions', 'sensa', and 'sense data'.

our mother's knee what it means to refer to black and white and all the other colours!'[3]

'Oh yes,' you say, 'of course; but that is not what I mean. When we learnt at our mother's knee, I thought we were all looking at the same things and must see them in the same way. Since then, I've read some philosophy and neuroscience, and now I can't for the life of me understand how anyone else, even our mother, could know what I am seeing when I look at a black font. My experiences are inside me, inside my head, in my brain. No-one else can see them. I may see the world completely differently from the way you see it.'

I can hardly take this seriously: 'Oh dear, you've become quite a solipsist recently. I feel like that sometimes, too. Don't worry, it will pass. Let's have a curry together.'

This, of course, is just a little annoying: 'No, no, it won't pass. You really don't understand. I am serious. I am me. You are you. You can't know what I experience in my own private world.'

'*Your private world*? Where on earth is that?'

'Don't play games with me. You can't see inside my head.'

'Well, actually, there is a sense in which I can. I can record from your neurons. I can scan your brain for blood flow changes, and many other things. You could do the same for me. We would find much the same stuff inside both our heads.'

'Yes, I know that. I don't think I am different because I am made differently—though equally obviously we are not physically identical. It's just that . . . well, I am me, and you are you. Don't you see?'

'Yes, I certainly see that, but I don't see why that entitles you to say that you have a private world that I can't know about.'

'Come off it! Obviously I am not referring to my neurons, or my blood flow changes, or anything else of a physical nature. I am talking about my experiences. I have my own sensory experiences, and you have yours. You know, they even have a name now. People call them qualia. You must also have them. Just look at one of those letters on the page. There is a black on white quale there!'

---

[3] Actually, this move is not as naïve as it may initially appear—see the denouement of this story.

'So you have become a dualist? You think there is something there that is not physical?'

'Oh no, not at all! These things are created by my neuronal processes, perhaps in a sense they *are* my neuronal processes—or at least what it feels like to have them. We are not going back to Cartesian dualism. I am not supposing a soul that interacts with my brain. In fact, as far as I'm concerned, I am my brain. And my brain creates these experiences that I see, feel, hear.'[4]

'I thought that these experiences were created in the same world that both you and I live in. That is why I was so puzzled at your reference to your "private world".'

'Well, yes, that's sort of true. But I am not referring to the things in the world itself. I am referring to the *quality* of the sensations I have when I see the world. That's why they are called qualia.'

'So, wait a minute. When you look at a black font, you think that there is not only the black font itself but also something else that is inside your head?'

'Yes, you've got it. I wish I could have put it that way myself.'

'But that's just another form of dualism. Why do you need to suppose that there is anything inside your head other than the neuronal processes that occur when you see a black font?'

'No, wait a minute. I don't think these qualia are a different sort of substance, something ethereal and ghostly.'

'But that's just what it sounds like to me! Tell me this. You are a scientist. You think that your brain is a material thing, though fiendishly complex. What possible experiment can you or I perform that will confirm whether or not these things that you call qualia exist?'

. . . long silence . . .

'Well?'

---

[4] This stage of the argument could have been the departure point for another version of the private language puzzle. The protagonist, ME, could have asked what on earth the 'I' was doing in this sentence. It would take another version of the dialogue to explore the problems created by this way of speaking about the relationship between the self and the brain, which we will explore later. This illustrates the point that there is a set of puzzles here that are all interrelated by various versions of the private language argument.

'Well, it's not like that. As I said before, you can't know what I experience, so I can't tell you.'

'So, what do you do? Do you, as it were, tell yourself—perform a kind of self-talk? How do you compare your own experiences?'

'Of course, that's easy. I know what I mean when I refer to black and I can remember what it was like. So, in a sense, I can tell myself "this is black".'

'But you can't tell me?! You must have a private language.'

'Well, if you want to put it that way, I suppose I do. But everyone does.'

. . . different long silence . . .

'Don't you?'

'Well, I am not sure about this. Tell me, where did you learn this "private language"? Does it have the same words as our language?'

'Well, I hadn't thought much about that. Yes, I suppose it does. At least, when I tell myself that I see black, I don't invent a new word. I don't refer to it as "kcalb", for example. Actually (Oh dear, this is getting muddling!), I don't think I use any words at all—I certainly don't need to.'

'So, this language is not a different language, and, possibly, it isn't even a language at all?'

'Well it's certainly not a language as we learnt at our mother's knee. But, look, it's very simple really. I see black. I know I see black. I remind myself that this is the same kind of thing—of quale, say—that I have experienced before. If I must put it into words, I suppose I would tell myself "I am seeing black."'

'And when you say that to yourself, you also tell yourself that you are communicating something different from what you would communicate to me when you say to me "I see black?"'

'Yes.'

'So what is different?'

'I've already told you. In one case I am talking about the font itself, in the other about how I see it—the quale that is my experience of the font, if you like.'

'But we don't know that these qualia exist. We have no way of conducting an experiment to see whether they exist. So why refer to them at all? Why not go back to when we learnt language at our mother's knee? When we saw black, mother said "that is black", so we learnt that that is what this colour is called. Isn't that simple? Moreover, we could only do that because all three of us looked at the same pictures in the same book. That's how we came to use the same language. If we had been French we would have said "noir", if Japanese we would have said "kuro", but we would still have ended up being able to do all that anyone ever can do to communicate what they see when they see black. Similarly, that is how we came to understand what people are referring to when they see red, green, blue, and all the other colours that we distinguish in our language. Mother didn't ask us whether we could see any qualia!'

'No, but she didn't know any science.'

'Hey wait a minute. This is not science! We seem to agree that what you are talking about is something—qualia, sense impressions, or whatever you want to call them—for which we can have no experimental evidence; that you communicate with yourself about seeing these qualia in a language that is either not a language, or just the same language you use to tell me "I see black"; that we are not talking about there being anything inside your head other than material substances of the same kind that I have in my head; so where is the science in all of this "private language" stuff?'

'OK. I agree that I am expressing a particular philosophical view of the world. But I also think that there are some kinds of philosophical beliefs that are necessary for us to conduct scientific investigation. How could I possibly study the brain–mind problem if I didn't think that something—qualia or whatever—was formed as a result of neuronal processes? Goodness me, this is the greatest challenge for neuroscience! You can't convince me that this is all a wild goose chase!'

'Maybe not, but if I am right then you don't even have a problem. In the sense we are talking about here, there is no mind–brain problem.'

'What?!'

'Well, it's up to you. If you think there is a problem, then there is a problem, but perhaps the problem is the way you are thinking, not a problem for science. I mean, look: you are proposing to investigate a phenomenon for which we can have no experimental evidence; that requires a language that is private (and I thought languages are for communication! . . . silly me!). Your line seems to lead to a modern form of dualism which I would suggest is alien to what science can seek to know about us as human beings.'

'I think we had better have that curry. You will have to tell me what *you* see as the relation between mental and physical events!'

At Aziz' restaurant

Much of this little book was conceived during wonderful Indian curries. Let's continue the discussion at the restaurant over a fragrant lamb kursi,[5] but we can drop the dialogue form.

Physiologists study the body. That includes the nervous system and the brain. Those particular objects of study have a special place in our world. Yet the proteins and other molecules that form the brain are hardly different from those we find in the rest of the body. Most often, they are identical. And the differences, when they occur, are not particularly mysterious.

Take the process of exchanging sodium and calcium ions. There is a protein that is responsible for this in most cells. It is commonly referred to as *ncx* (which stands for the sodium–calcium exchange). In the eye, this is replaced by a related protein that also transports potassium ions in the photoreceptors. There is a logic to this. For photoreceptors to work properly, the relevant cells must have very little calcium in them. So calcium has to be drawn out of those cells, or, to put it another way, it is necessary to generate large calcium gradients. The change of protein helps to produce this effect.

[5] Like the bistro omelette in Chapter 2, a whole leg of mutton has an important message for the integrationist–reductionist debate. The same spices, the same yoghurt, onions, garlic, oils, and meat are used to prepare a whole leg of lamb or mutton as are used in other curry dishes. But the final result using the whole leg cannot be compared with an ordinary curry prepared with cut-up meat pieces. It cooks as an integrated whole and it produces a remarkably different culinary experience from the 'reductionist' version.

Of course, a separate gene codes for this special version of the *ncx* protein. Similarly, the sodium channels that occur in the brain and in the heart are different (or at least, the expression levels of the various channel proteins are different) and they are coded for by different genes. When neurophysiologists and other neuroscientists refer to the special features of the brain and nervous system, however, they don't usually have these kinds of technical molecular differences in mind. What they are referring to is the problem of consciousness.

There are many ways of identifying this. We can speak of a mind–brain problem, and of related questions that either overlap strongly with the problems traditionally regarded as within the field of philosophy, or even, on some accounts, *are* those problems. This is a contentious area. Neuroscience is viewed by some as coming along to solve the great philosophical mysteries of man. Why are we conscious? Is there a self? Do we have free will? It is not only neuroscientists who are fascinated by these questions. Molecular biologists, quantum physicists, cosmologists, mathematicians have all had a go at dealing with the problems of consciousness and the mind. A list of all the titles that have tried to do this would fill a volume as long as this.

To suggest, therefore, as the opening dialogue of this chapter does, that these problems do not really exist or that, if they do, they are very different from what many writers have portrayed them to be, may appear shocking. To some of my readers, perhaps even more shocking than the mathematical surprises I have presented in previous chapters. Bear with me.

We are conscious. We cannot be so without a functioning brain. Therefore, we locate consciousness in the brain. The brain is seen as a performance space, in which the drama of the body is presented. It is as though there is an 'I' (appropriately pronounced like 'eye' in English) that 'looks' at the end results of complicated neural processing. This is what some people call the Cartesian theatre, derived as it is from Descartes' dualist philosophy of brain and mind.

These results are viewed as 'maps' of the world outside. On this view, 'we' do not directly see the world. Instead, 'we' ('I') interpret these maps so that we have the illusion of seeing the world directly.

But in fact what 'we' 'see', 'hear', 'smell', etc. are representations of some sort—much like 'qualia', in fact.

Whether people articulate it in this way or not, the assumption is that there are visual qualia, auditory qualia, and so on. The 'brain theatre' is therefore a 'total sensurround experience!'—the grandest of opera theatres. So, if I listen to Schubert's piano trio, my experience can be resolved into a succession of auditory qualia—or perhaps a single quale (I have never understood how we are supposed to divide this stuff up). The real events, by contrast, are, on this view, the sound waves in my room, as they would be measured by a physicist with complex instruments. The Silmans were right after all!

There are at least three problems here. First, where (or what) is the 'I'? Second, where and what are the maps of the external world? Third, will these entities stand up to critical analysis any better than qualia?

The traditional Western philosophical view of these questions is clear. It supposes that there is an 'I', the self, the mind, the soul. This is separate from the brain even if dependent on it. This view was inherited from philosophers such as Descartes by people like Charles Sherrington and John Eccles, distinguished neurophysiologists of the twentieth century.

More recently, neuroscientists have rejected this view in favour of various forms of 'the brain is the self'. The Cartesian view makes sense only in terms of some mind–brain interaction. But that is problematic. Everything would be much easier if it were not needed. So people are naturally tempted to cut the Gordian knot and simply say that, in some sense, the neural activity of the brain just *is* the 'self'.

If it is complicated enough, they say, then this neural activity 'produces' consciousness, including our sensory perceptions. The secretion of insulin is a special property of the pancreas, pacemaker rhythm is a special property of the heart, and likewise consciousness is a special property of the relevant neural networks. This is not a trivial claim. It has major consequences. People think, for instance, that if they could super-freeze their brains when they die they might eventually be resurrected as the same person. This way of thinking also leads to all kinds of conundrums involving computer simulation.

Could a computer simulate consciousness? What would it have to do to 'be conscious'?

This approach represents something of a modern consensus amongst neuroscientists, and some philosophers. There are difficulties with it. How to formulate these ideas without falling into some version of the 'private language' argument? I don't think the concepts stand up any better than do qualia.

It would require a book of enormous length to unpick these difficulties fully. These problems are notoriously difficult to unravel because they are so deeply ingrained in our language. We say 'Use your brains!' when we just mean to say 'Think!' Fortunately, I don't have to write such a book. Others have already written much of it (Armstrong 1961; Bennett and Hacker 2003; Cornman 1975). But it will make sense to sum up the essential points. The story format will serve to that end.

Action and will: a physiologist and a philosopher experiment

Suppose we accept the idea that the 'self' is, or is in, the brain. We then face the question 'Where is the "I"?' One answer is: it is wherever in the brain intentions are generated.[6] Could we imagine finding that spot? Or could we even identify the networks that are responsible, which could be quite diffuse? Can 'I' be identified as a network of neurons? Let's follow this idea through with a specific example.

Here I am, writing this book. Then I am interrupted. Someone comes in and asks where something is. I am engrossed in writing, so I do not reply in words. Instead, I point to the object in the room. This is an example of an intentional act. Imagine, then, a physiologist. He studies the muscles and the neural connections involved in this act: the raising of the hand and the pointing of the finger. He arrives at

---

[6] If the reader has absorbed the underlying ideas of this book, the alarm bells will already be ringing. To talk of intentions being 'generated' by the brain is already to fall foul of the traps that our language has in store for us through centuries of thinking in particular ways about mind–brain relationships.

what he sees as a completely sufficient explanation of all the movements involved.

We might imagine that this analysis is achieved in stages. First, it is necessary to understand in a general way the mechanisms involved in raising an arm. Then this understanding can be refined to cover the specific movements that form a prelude to pointing at something. Finally, the physiologist can analyse the finer movements involved in the pointing.

Going on from there, he may even find that the neurophysiological apparatus concerned can all be triggered by stimulating a small region of the nervous system, a particular location in the motor cortex, say. Or that it can be achieved by stimulating a finite number of points in a patterned sequence. By this stage he is no doubt feeling quite pleased with himself. He is ready to publish a paper on 'The neurophysiological basis of pointing behaviour'.

When he does, a philosopher reads it. She notices what she considers to be at least an example of loose terminology and at worst a complete misunderstanding of a major distinction. The paper concludes: 'We now have a complete physiological explanation of the action of pointing.' It is the use of the word 'action' that triggers her reaction. When she meets the physiologist later, she tells him of it. 'You have not explained the action of pointing', she says. 'What you have explained is the neurophysiological basis of a particular movement or set of movements. You have progressively refined that set until it corresponds outwardly to what we see happen during the action of pointing. But you haven't provided an explanation of that movement *as an action.*'

The physiologist is taken aback. He has vaguely heard philosophers use this sort of distinction. But he has always thought of it as something to do with the way we use everyday language in referring to behaviour. Surely it can have little relevance to whether he has traced the correct neural pathways generating the movements concerned.

His position is a tentative reductionist one: tentative because his approach is to see how far he can get with his explanations purely in terms of neurophysiological events. And if, he says, he can satisfactorily account for all the visible movements during the action of pointing then surely there is nothing else left to explain. Or is the philosopher postulating the existence of different, mysterious forces that operate only when the movement, the *same* movement he insists, is said to be the action of pointing?

Surely not, he thinks. For if there were no evidence for these forces other than their visible effects in the movements of the muscles concerned, then that would be to advance a hypothesis that cannot be tested. And even if the idea were correct, how do we know that these forces and any mental events that may be said to accompany them are not mere epiphenomena—just a sideshow to the real events, the neuroscience? He has in fact a suspicion that consciousness and mental events in general may be in this category.

The philosopher replies that she starts from an undeniable fact. In common discourse, and in the whole set of presuppositions about behaviour and responsibility that go with it, we need to make a distinction between movement and action, and we do. Only if the movements concerned occur within a context could it be said that a particular *action* occurred. That context is defined by the conditions necessary for ascribing an *act* to a *subject*.

Thus, it might be that the subject intended to point. Not all actions require intentions (one can act unintentionally) but it would certainly count as a criterion for a movement to be described as an action for it to be intended. The philosopher offers to demonstrate this point. All she asks is that the physiologist accepts to be the subject of the experiment. Only then, she explains, will it be possible for him to see if the experiment has worked.

So, under directions from the physiologist (and we suppose there are no technical difficulties in doing this) she arranges electrodes in the appropriate places and then stimulates in the way proposed in the article. The physiologist's arm moves and his hand points. The philosopher asks him to say what has happened.

He replies, 'Yes, I see. I felt that this movement was, as it were, imposed on me. It did not differ externally from an act of pointing, and yet *I* did not point. So, I will grant you that my electro-physiological stimulation pattern does not reproduce *all* the brain states that may be needed. But, if that is the basis of your argument, then it is only a delaying tactic. If you give me time, I'll work out the full picture. I'll identify the neural pathways that were excited before the set of neural states which my stimulation pattern mimicked. I take it that you don't dispute that there will be some such neural activities. Eventually, I must find the pattern that not only produces the movements concerned but also avoids giving me the feeling that I was forced to move. If I succeed, then I will have located the neural basis of "me", the "I" that does the intending.'[7]

Now it is the philosopher's turn to look puzzled. She realises first of all that she has made a mistake, even from her own viewpoint, by presenting this issue as though it were an empirical one to be determined experimentally. The point is that in practice we *know* when we act intentionally. We don't need to study our brain states to know that. So, the protagonists in this story leave each other puzzled.

Explanatory shift between levels

Within the context of that story, it is difficult to resolve the argument. It seems that the physiologist must be right in thinking that he can trace neural networks and their activities back as far as he wishes, and that he must eventually find a pattern that 'produces' the intentional action, that is, does so without giving the agent the feeling that 'he' did not act.

We are assuming here that our philosopher doesn't deny that, when we act intentionally, some neuronal circuitry must be activated. She is not a Cartesian dualist. So, it may be fiendishly difficult to determine the pattern required, and it may be even more difficult to mimic it, but these seem to be technical problems not conceptual ones. But, equally certainly, our philosopher was correct in realising that her experimental test was misleading. It succeeded in its limited aim to show that reconstructing a movement is not necessarily

---

[7] This is not far-fetched. Experiments of this kind have in fact been performed to try to identify neuronal activity that precedes an intentional act.

reconstructing an action. But it gave the impression that the issue at stake, the distinction between actions and movements, is an empirical one. It is not. It is conceptual.

We like to think we act intentionally and rationally, at least some of the time. Do we? How to decide? Is this a purely empirical question? Surely not! The argument is basically very simple. We cannot, coherently, deny our own rationality. Otherwise how could we mean what we say or be convincing in saying it? That is precisely what happens in the sad cases of those mentally ill people who nevertheless are aware of, but can't help, their irrationality.

Suppose we really could succeed in 'reducing' rational behaviour to molecular or cellular causation. In that case, we would no longer be able meaningfully to express the truth of what we had succeeded in doing. In any event, the question does not arise. No such reduction is conceivable. We know what it is to be rational, and what it is to lose that capacity. That knowledge has nothing to do with the question whether there exist specific and causally sufficient neural states and interactions while I am writing this book, for example. Of course there are. So what?

If we could discover them, they may well provide a complete explanation for how my brain operates while thinking and writing. But they would not lead to the discovery of where 'I' am to be found. Nor do 'I' need to consult my brain states to know what I am doing and intend to do.

This illustrates the main claim of anti-reductionism in science. Such a complete explanation of mechanisms at one level does not necessarily explain what exists and happens at higher levels. Indeed, we may need to know about the higher levels in order to explain the lower-level data that form an input to the mechanisms involved. That is one of the lessons of Chapters 4 and 5. How can we apply those lessons to the present problem?

The mistake our physiologist makes is to think that when he gets down to tracing back all the causal interactions that underlie and precede my activity of pointing, he will be working purely inside the brain. To see and understand that mistake we have only to ask what it was that I was pointing at.

The dog needed to go for a walk. Someone wanted to take him out. Only, where was the dog lead? That was the context.

Obviously, any explanation of my pointing *as an action* would need to take that social context into account. This would include the semantic context in which it makes sense to say of my action that it was 'showing where the dog lead was to be found' (what linguists call the semantic frame). Given that context, on the other hand, the question of how to explain my action is trivial. Not only did I and the person who questioned me understand it immediately, so did the dog, who immediately recognised the prospect of 'walkies'.

But won't all that also be 'represented' neurally? Won't there be maps inside my brain that 'show' all of these interactions and contexts? And surely it must be possible for the neurophysiologist to find these? It is instructive to first grant this idea and see where it leads. Let's suppose that there are indeed such maps, such representations of the outside world and my social context within my brain. Who or what on earth then consults these representations?

The answer might seem obvious. 'I' do. But remember that we were hoping (or at least our physiologist in the story was hoping) that we would end up identifying the 'I'. Instead, what we find with this way of thinking is that the 'I' is continually pushed back, as it were. We will never find it by tracing neural networks. We would always have to suppose that there must be another part of the nervous system that 'looks' at these representations of the outside world.

As the epigraph at the head of this chapter shows, the temptation to look for such a location is very strong. Anyone who thinks that consciousness is a property of a part of the brain must follow Francis Crick in seeking to identify it. But this is a very strange endeavour altogether. I suggest that there is, and can be, no such division of my brain and nervous system. What is happening here is that we are tying ourselves up in philosophical knots. We are puzzling over entities, like qualia, that seem to exist only as a result of the philosophical conditioning we are subject to. We are confusing ourselves.

It is remarkable how deep-rooted these confusions are. We are profoundly attached to the notion that somewhere in the brain there must be representations of the outside world[8] and that somewhere else there must also be a part of the nervous system which can be identified as the 'me' that does the 'looking' at these representations. This is the alternative version of the private-language argument I referred to earlier. The 'I' in this sense is just as unnecessary as qualia, and reference to it creates much the same kinds of philosophical puzzle.

The self is not a neural object

We can't divide the brain up and say that there is a part of it which is the 'theatre' representing the outside world for 'me' and another part that is 'me' looking at these representations.[9] This way of thinking confuses levels at which different entities can be said to exist.

The problem is not just that this leads to an infinite regression. It is also that 'I' or 'me' or 'you' are not entities at the same level as the brain. They are not objects in the same sense that the brain is an object. In the sense in which my neurons are objects, in which my brain is an object, and in which all the other parts of my body are objects, 'I' am nowhere to be found. This does not mean that 'I' am not somewhere. While I am alive I am obviously where my whole body is. If my body is in England then I am in England. If my body is in France then I am in France.

---

[8] It is important to understand that representations or maps in the sense used here would have to be more than just a database of connections formed as a result of repetitive neural activity. The essential feature of a map is that some-one can read it as such. No-one, not even me, reads the states of my neurons in this way. If I play a guitar piece, for example, there are certainly neuronal circuits active that enable me to play with dexterity, but these are not maps of me playing the guitar. They, and the finger movements involved, *are* me playing the guitar.

[9] Of course, there are many ways in which we can functionally divide the brain up from a physiological point of view, for example into sensory, motor, and association areas. The study of brain lesions, split brains, and brain scanning enables even further subdivisions of function. I am not arguing against a *physiological* attribution of functions to various parts of the brain. The argument here specifically refers to identifying the 'self' with the brain or a part of the brain.

This is a problem of levels and ontology. We have encountered this kind of thing in Chapter 5. There we saw that it did not make sense to say that cardiac pacemaker rhythm exists below the cellular level. There are no molecular oscillators that generate it. Instead, an integrative activity emerges from a multiplicity of protein interactions at the cellular level.

We cannot locate the site of pacemaker rhythm at the sub-cellular and molecular levels. Yet we have no difficulty in locating it at the level of certain cells within the whole organ. We know what it means to refer to the pacemaker of the heart and we can locate it anatomically. The fact that it cannot be found below the cellular level is irrelevant.

If a particular biological function or entity does not exist at one level that does not mean to say it does not exist at all. To identify it may be quite simple once we have made the requisite explanatory shift. We just have to go up or down a level or two to find the context in which that entity can be said to exist. One of the important goals of integrative systems biology is to identify the levels at which the various functions exist and operate. In the case of the brain, this may be very difficult to achieve. But it is nonetheless possible and necessary.

Now let's return to our neurophysiologist. Remember, we left him tracing back further neural connections and patterns of activity in the hope of finding the neurophysiological basis of the 'I' that intended the action. What he would find at that level would be the sort of 'conditions explosion' that we encountered in Chapter 4. In the case of my pointing at the dog lead, these would include the huge number of conditions (states of neurons and synapses, for example) that occur because of the context in which my action makes sense.

At the neural level, these will be an unexplained set. They will always remain so, at that level, even if we were to be successful in identifying them all. Explanation is possible only at the appropriate level, in this case the level at which it makes sense to talk about dog leads.

Moreover, even if the physiologist were able to uncover all the conditions that relate to my action, they would be mixed in together with many others that relate to the world outside and to the states of other people and, in this case, of my dog. And there would be no way of telling which conditions are which. That is what happens in a 'conditions explosion'. And when it happens, we know that a shift of explanatory level is required.

It makes no sense to say that all of these states are mapped in my brain. We might as well say that all the chemical properties relevant to the functioning of life must be coded in the genome. In both cases, the genome or the brain, we need to recognise that these are databases that the system as a whole uses. They are not programs that determine the behaviour of the system. And in both cases, nature will have been parsimonious in what is needed in the database. It will be a partial database, not an exhaustive one.

It is instructive to think of this in the context of the way in which a musician interacts with his library of sheet music. At one and the same time, the sheet music is less than what the musician has, and it is more.

It is less, because—like the omelette recipe (Chapter 3)—much of what the musician knows is not written explicitly in the scores. If all musicians were to die tomorrow and it took a few generations for humans to develop musical talent again, we would have great difficulty recreating the culture from the musical scores alone. Indeed, we already have this kind of difficulty when trying to recreate, for example, the medieval music of the troubadours from the very different musical notation they used.[10] It would be like recreating extinct organisms from their genomes alone, without a fully working maternal cell and the equivalent of a womb.

At the same time, there is more in the musical scores, because the whole library of a musician is not stored inside his brain. His skills and knowledge interact with his musical library to generate the full range of music he plays.

---

[10] A modern example would be the music of Japanese Noh plays. There is sufficient notation for experienced Noh artists to interpret, but they rely mostly on tradition, usually handed down through a few families of players. It takes 30 years for a Noh musician to reach his peak.

The deep-frozen brain

The idea that the person can be identified with his/her brain is now deep in our culture. I suspect that it will take a shock to shake people out of it. Nevertheless it is instructive to try.

Of course, to suffer damage to the brain is most likely to endanger the integrity of the self, whereas damage to other parts of the body at least leaves a 'self' still present. From this people infer that there must be a part of the brain that in some sense *is* the 'self'. If that part remains functional then so does the self. But there is a distinction between necessary and sufficient. The brain is clearly necessary for there to be a functioning self. Is it also sufficient?

Let's perform the complementary experiment. Suppose we progressively remove the other parts of the body. If we remove the eyes we have a blind person. If we remove the ears we have a deaf person. These are 'selves' that are seriously impaired by what we are imagining, but I think that most people would agree that the essential 'self' can be said to remain. If we remove both eyes and ears, we have a deaf–blind person. Again a self can still be identified, though greatly reduced in faculties.

Now remove the skin and all parts of the body that are essential to touch. We don't have examples of such people, of course, but would we still have a functioning 'self'? It begins to look unlikely. Consider what happens when people experience sensory deprivation. In a soundproof room, they float in a uniform temperature with no light. Very soon, they find the self begins to disintegrate.

Take this process even further and add paralysis to remove functioning limbs, including becoming dumb. So now we have a 'person' who is blind, deaf, mute, sensation-less, motionless, gutless—but 'possesses' a brain that still has a blood supply. This is a bag of neurons in a perfused pot. Do we really think that a viable 'self' would remain?

What is happening here? We are completely separating the brain from the body. At this point, the question of consciousness, or 'self', ceases to have a normal meaning. No conceivable experiment could answer any question about it. The brain on its own does not communicate. It doesn't make sense to ascribe consciousness to something that cannot communicate.

It's obvious, isn't it? Yet people put good money on the idea that, by deep-freezing one's brain at the time of death, one might hope to be resurrected in the future as the same person.

The resurrected self?

But surely 'I' could still be resurrected as 'me' if 'I' receive a body transplant. Or perhaps I mean that 'I' give a brain transplant to someone else? Whichever way you express it, surely 'I' follow my brain, not the other way round?

Let us examine this more closely. What would happen if we could reconnect a disconnected brain to a new body? Let us assume that all of this is technically possible—the deep-freezing, the unfreezing, the reconnecting. What would we end up with? Who would this 'resurrected' self be? Would it be the same self as before?

What would be the difference between this and replacing parts of the body with transplants or prostheses? Or, to reverse the question, what's the difference between a brain transplant and, say, a heart transplant? Won't the 'self' follow the brain, not the rest of the body?

I'm not sure we could answer that question definitively. Even without such a traumatic procedure, changes in body chemistry, for example during psychiatric drug treatment, can make even the patient ask whether he is still 'himself'. That can happen with a single drug. Here we are talking about thousands of 'drugs'. Since the body, and its blood chemistry—all those circulating hormones—would be different from the original one, the change in personality would be profound. People might well have great difficulty knowing how to relate to such a person.

To see why, try the following thought experiment. Aged 'mother' who died, let's say 25 years ago, comes back looking like a young girl with totally different features and grossly different personality because her new body has different hormonal balance and different gene expressions. Yet, she claims (correctly) that 'she' could remember what mother used to remember. I am not sure what we would want to say. Still-alive aged father would hardly know what to think or do! We might say something like 'You remember many things that happened to my mother but, apart from that, you don't seem to be my mother.' We might even say 'It is as though my mother's memories have been transplanted into you.' And I think we might feel that however much the person protested 'But it really is me!' What, precisely, does that statement mean? It is not as simple as asking where the (single, unitary) 'self' is or has gone. We might end up saying that bits of it are here, other bits are there, many bits have been completely transformed, and yet other bits no longer exist.

Of course, this is just a thought experiment, almost in the realm of science fiction. But we can get glimpses of the implications from our own experience. I suspect that most of us, if we are honest with ourselves, have had the disorienting experience of having to ask who we are and where we are. These are moments, sometimes when waking from dreams, when the self appears to come apart before it clicks together again. Some forms of meditation have this as an aim: to be able to deconstruct and reconstruct the self at will, so as no longer to be subject to the disadvantages associated with it (greed, anger, etc.).

My point is that the 'self' is an integrative construct, occasionally a fragile one. It is also a necessary construct. It is one of the greatest symphonies of the music of life. But if we play around with its physiological basis as envisaged in this chapter, we might have to re-examine some of the fundamental ways in which our language works.

These thought experiments are intended to shock. Philosophically speaking, they suffer from a problem similar to that which the philosopher faced in experimenting on the physiologist a few pages back. There is a danger of making the issue appear to be simply an empirical one. Even so, an empirical experiment (real or imaginary) can sometimes help to convey a conceptual point. This is particularly true of situations where the distinction between the empirical and the conceptual is difficult to keep in mind. It is when we are forced to reconsider our language that we can easily tie ourselves in linguistic knots.

At the level of neurons and parts of the brain, what we normally mean by the self, that is, you or me, is more like a process than an object. We can certainly ask questions about how you or I would be affected by damage or alterations to our brains, that is, how the integrity of the self is compromised. And the impact on the self will, of course, be different depending on which part of the brain is damaged. But when we start to talk about the location of the self, we are talking about a person. Such talk belongs to a context in which it makes sense to refer to persons. It leads to semantic confusions to recast these as questions about locations in the brain.

The symphony of life that we call a person is not just the playing of individual instruments in the orchestra. And there is no Cartesian opera theatre in the brain.

10 § Curtain Call: the Artist Disappears

Each beat and each tune indescribably profound, no words are
needed for those who understand music.

Zen Buddhist parable

Jupitereans

Suppose that we found life similar to ours on, say, one of the moons
of Jupiter. The first humans to travel to this new world report back to
earth the message that, against all expectations, the space travellers had
found 'God' up there. For the people they find have what, to all
intents and purposes, are cathedrals. There are 'priests' dressed in
colourful robes. There are ceremonies of great spiritual importance
for the big events in life and death. They practise a special form of
'prayer'—meditation—that is of great psychological value. The space
travellers have sophisticated physiological instruments with which to
measure the contentment and reduction of suffering. Moreover, there
are scriptures! Masses of them.

A high-level deputation of presidents, kings, and church leaders
gathers on earth to decide what message to send to the travellers.
They all focus on the one important next step. As quickly as possible,
learn to read these scriptures, learn enough of the language to discuss
them with these people!

As the space travellers progress in this endeavour, the messages they
send back to earth become increasingly enthusiastic: 'These people,
particularly the leading "priests", really do think like us. We have
already established the key words used in religious contexts, and they

map almost exactly on to ours.' Their enthusiasm is more than matched by that on earth. Bishops, rabbis, and ayatollahs are scrambling over each other to make suggestions for how the travellers can dig deeper and, perhaps, settle once and for all the differences between their theologies. They encourage the travellers to set up medieval-style debates with the spiritual leaders of the new world.

Inevitably, one of the travellers outpaces the others in mastering the language. She starts to ask the right, penetrating questions. In advance of a set-piece public debate she spends hours of time each day with a few of the Jupitereans. But something strange happens. The more penetrating the questions she asks, the more they reply in almost enigmatic terms. She starts to have doubts. It is almost as though the Jupitereans are laughing at the questions. They don't answer them, they simply ask in reply 'Why do you need to know the answer to such a question?' Good Oxford philosopher as she is, she sees traces of Wittgenstein in what she hears.

So, becoming anxious (the day of the grand debate is near and she is supposed to advise her leaders on how to conduct it) she tries another tack. Instead of asking philosophical questions, she seeks to establish whether all the words relating to religion, to which they had automatically assigned earth-like meanings, really do carry those meanings.

She starts with 'the soul'—or what the non-religious on earth would call 'the self'. And she finds that, in the earth sense, it doesn't exist! It seems to be a process rather than a thing. 'Everything is in a constant state of change', they say. They seem to speak in verbs. The nouns are often not there.

So she moves on to 'God'. She tells them about Newton and Einstein, about our understanding of the physics of the universe, and finds strict correspondence in the Jupiterean science. She explains that many earth people think that there is something—called God—that holds all of this together; ensures that the laws are just right for the evolution of life, etc. At this point she comes to realise that they just don't have the concept.

Things just 'are'. They don't need a creator. There is no 'God' as a person, and the founder of their religion is not a God. So, the space travellers made a big mistake when they initially assigned the 'god' word in the language to the concept of a creator. She now realises that it has little more significance than the integrated 'spirit' or 'essence' of something.[1] She comes to understand this as she finds that everything, even a stone, can be said to have such a god.

Now she really panics. She realises that what she has discovered is a big hole at the centre of what many on earth are expecting to be a major revelation, but as a scientist she is delighted. 'At last', she says, 'I have found a treatment of the spiritual that makes sense!'

Well, we didn't need to wait for space travellers to go to the moons of Jupiter to have this experience. This is almost exactly the discovery that the Western missionaries made when they first encountered Buddhism.[2]

Role of culture in our view of the self and the brain

Many of our misconceptions about mind and brain are ingrained in our language, and have been so for centuries. So it may help to step outside our own language and culture to try and see how differently we could conceive the world in terms of another language. For different cultures can conceptualise factors like 'mind', 'soul', and 'self' (and indeed theological concepts like 'god') quite differently. Of course, there is no privileged language or culture. Other languages than our own conceal linguistically based illusions. But these are likely to be very different from our own.

All languages are prisons of culture as well as liberators of communication. We need language to communicate, but our languages in turn cloud what we understand. There is no mystical oriental culture in which all our problems are solved. The point is rather that cross-cultural experience can help us break out of our illusions.

It is instructive to see how a 'god-less' and 'self-less' religion teaches its practitioners the path to spiritual experience. Here is a

[1] The 'god' word in Chinese, Japanese, and Korean, 神, works rather like this.
[2] The debt to Stephen Batchelor will be obvious to those who know his marvellous book, *The awakening of the West* (Batchelor 1994).

story from Zen practice. I have chosen it because this tradition is one of the least overlaid with metaphysics. It can therefore speak more directly to a largely secularised society.

It is about an Oxherder, a boy who doesn't know where his ox is. Traditionally it is told in ten illustrated poems:[3]

The Oxherder

- He aimlessly prods amongst the grass as his path becomes longer and the mountains more distant. He is exhausted and dispirited.
- Nevertheless, by the water and under the trees there are traces of the ox. He asks the long fragrant grasses 'Did you see the ox?' He wonders how the upturned ox horns could possibly be concealed.
- A bird sings from a branch while the sun is warm, there are soft breezes. Beyond the green willows on the bank there are the unmistakable ox horns.
- He catches the ox, but the ox is strong and cannot be tamed easily. Sometimes he charges off up to the high plateau, deep in the mist, and refuses to leave.
- He doesn't let go of the whip or the rope. The ox becomes gentle. Even without the rope it would now follow the boy.
- The boy rides the ox, playing flute melodies that echo through the sunset clouds. Each beat and each tune indescribably profound, no words are needed for those who understand music.
- He has ridden the ox home, there is now no ox there and he is at his ease. The sun is high and he is still dreamy. The whip and the rope are abandoned in the thatched hut.
- Whip, rope, man, and ox—all are non-existent. The blue sky is vast; no message can be heard, just as the snowflake cannot last in the flaming red furnace. In this state, one can join the ancient teachers.
- In returning to the fundamentals, going back to the source, I had to work so hard. Perhaps it would be better to be blind and deaf. Being in the hut I do not see what is outside—the river flowing tranquilly, the flower simply being red.
- He enters the city barefoot, with chest exposed. Covered in dust and ashes, smiling broadly. No need for the magic powers of the gods and immortals, just let the dead tree bloom again.

---

[3] The version I give here is loosely based on Wada's *The oxherder* (Wada 2002).

The story is used as a guide to the meditative process, by which one can subdue the mind and ultimately 'forget' the illusions of self, as part of the process of enlightenment. The idea of forgetting in this sense is not unique to Buddhism. Pre-Buddhist Chinese philosophers expressed similar ideas. There is a Taoist philosophy of 'forgetting'[4] that I find particularly relevant to the practice of music. A skilled musician literally 'forgets' in the sense that he does not think about the deliberations that were involved in learning a piece of music. In effect, he just says 'go' and it happens. Paradoxically, the more he 'forgets' in this sense the more he is in control of what he is playing.

This is a highly refined process, and the beauty of it is that the musician himself can also, as it were, enjoy observing himself play. This is quite a good analogy to what the 'self-less-ness' or 'self-detachment' of meditation is seeking to achieve. Buddhists call it 'letting go'. Some musicians practise meditation precisely to encourage this process when they perform.

It is much easier to understand what is happening here if one regards the self as an integrative process that can be deconstructed rather than as a neurological object.

If the 'self', the 'I' as Descartes conceived it, and in new guise as modern neuroscience tends to conceive it, is an object that we hang on to because our language and culture make it very difficult to do otherwise, then clearly it is important to know that there are cultures in which it doesn't exist (Houshmand *et al.* 1999)—or at least in which it doesn't exist in these senses, neither as a separate substance (the Cartesian view) interacting with the brain, or (the modern view) as part of the brain itself.

For 2500 years this has been part of the aim of Buddhist meditation. There are many forms of Buddhism around the world, with a wide range of practices and beliefs, but the idea of 'selflessness', 'the disappearing self', and 'letting go' is common. In some forms, there is little or no metaphysics either, just a code of practice: a religion, one might say, without beliefs. And, as such, it contains no possibility of conflict with science.

---

[4] The best account I know of this philosophy is by Jean Levi (Levi 2003).

My purpose in ending my book with this brief chapter is not, however, to promulgate Buddhism. One can appreciate an insight from wherever it may come, whether or not one agrees with the rest of the package in which the insight is to be found. So, just to reassure anyone who may be worried by my reference to an oriental religion, there have been Christian mystics, notably Meister Eckhart, who expressed some of the same insights, although I suspect that it must have been vastly more difficult for him to do so within his culture than it has been for Buddhists in Eastern cultures. There are very few followers of Eckhart today in the Christian tradition. Millions follow the Buddha.

One reason may be that the possibility of 'absence of self' is deeply ingrained in some of the languages of East Asia in which Buddhism flourishes. If Descartes had been Japanese or Korean I think he would have found it quite difficult to formulate his famous 'cogito ergo sum'—'I think therefore I am'. The most natural way of saying the Japanese or Korean equivalent would be 'thinking, therefore being'. The subject is not usually there.[5] The words for 'I' or 'me', and even more so the word for 'you', are only used for emphasis.

Nor does the word for 'I' creep back, as it were, in the verb form since, unlike Latin or English, the verb form does not conjugate with the subject. 'I am', 'you are', 'he is' are all the same. The context or, if that will not suffice, a reference to a person's name, is what tells us who is concerned.

It seems to me that what these languages do is to emphasise the 'doing-ness' of things, the processes that occur, that is, the verb, rather than the subject who is the possessor of the being-ness or doing-ness. Often, the verb alone is the complete sentence, as though it doesn't need anyone to possess it.

Not surprisingly, therefore, the concept of the 'self' in such a culture can have much more in common with a process than a thing. When trying to escape from the confines of my own culture and language I find it very helpful to think of the self in this way. So, the

[5] Of course, Korean and Japanese *translations* of 'cogito ergo sum' include the 'I', as they must to make any sense of what Descartes was saying.

self, the 'me', is where my body is (Chapter 9) because it is one of the most important integrative *processes* of my body. *Systems* biology indeed!

Thinking in this way, I am more likely to avoid the philosophical maze in which it is so easy to get lost in a 'private language' dead end. I don't need to think of 'I' as an object, so I don't need to find a part of the brain in which it is located.

This may initially seem strange to Western eyes and ears. It did so to me initially, but then I found that the more I used East Asian languages like Korean and Japanese, the less awkward, the more natural, the dropping of the subject seems to be.[6] One's mind focuses on what is happening, on processes, on what is being done. The absence of an explicit identification of the subject achieves that, and so it structures one's thinking. Viewing the self as a process rather than an object then becomes more natural.

The self as metaphor

This book has used metaphors and metaphorical stories. The aim has been to stimulate changes in mindset that seem important. Already I can hear the chorus of criticism from some scientific colleagues. 'Can't you say everything you want to say in literal, scientific, language?' The short answer to that is 'No!' A slightly longer answer would be 'No, *and nor can you!*'

Metaphor lies deeper in our languages and our thought processes than we might wish to recognise. There is very little that we can say of great human significance without metaphor. This is obviously true of poetry and other literary forms. It is also true of scientific language.

Even to say something as scientific as 'The temperature is high today' is already to use a metaphor. Why should heat be measured by metaphorical reference to the dimension of height? We do so because a mercury thermometer rises when it is hot. What if our first thermometers had been bimetal strips that bend downwards when hot? Then we might in summer say 'The temperature is down today.'

---

[6] It is for this reason that such languages are called 'pro-drop' languages.

In biological systems, too, we speak of lower and higher levels. Genes and proteins are at the bottom, organs and physiological systems at the top. Biological scientists would find it hard to manage without such a classification of levels of organisation, but we should recognise that this also is metaphorical. Genes are all over the body in every cell; so is the nervous system: it ramifies everywhere. These 'bottoms' and 'tops' of the biological scale are figurative.

Every metaphor produces its own forms of prejudice. In this case there is a feeling that the higher levels need to be explained in terms of the lower levels. The reverse does not seem so natural to our scientific mindset.

But we don't own nature. Nature does not need to share our mindset. In truth, it has no mindset and will serendipitously exploit any of the interactions that it finds are functional and can add to the chances of survival. It is likely, for example, that in the early stages of evolution, in the so-called RNA world, the distinction between genes and enzymes did not exist.

Words like 'high' and 'low', 'in' and 'out', 'up' and 'down', are frequently used in language in such metaphorical ways. We can't do without them. Such metaphors have been in our languages so long, dead and buried like fossils in the rock, that we have ceased to think of them as metaphors. Therein lie many of the philosophical traps that our languages have in store for us. Precisely because we are unaware of the way they predispose our thinking, we find it more difficult to escape from it.

The self is also such a hidden metaphor. It is a very useful and important one, too. We might say that it is 'as though' a self, as a virtual object, is doing all the things that 'I' do. We need such a metaphor for many other aspects of our culture to fall into place. For example, we need for legal reasons to attribute responsibility to people.

But none of these cultural needs require that a physical object, rather than a coherent integrated process carrying the characteristics of an agent, should have evolved. We have no difficulty in assigning legal responsibility to other things, such as governments and companies, that are not physical objects, but which clearly are agents. In relation to a self, it is the coherence and rationality that matter, not whether there is a bunch of neurons with which 'I' can be identified. The reason we can naturally think of the self as an object is that it is always associated with a particular body.

The artist disappears

I gave the title 'The Music of Life' to this book because music also is a process, not a thing. And it has to be appreciated as a whole. It is notoriously difficult to describe in words. As the Oxherder comes to realise,

no words are needed for those who understand music

or if you prefer a Western philosopher saying much the same thing, you can use the ending of Wittgenstein's *Tractatus logico-philosophicus*:

Wovon man nicht sprechen kann, darüber muss man schweigen
[that whereof one cannot speak, one must be silent].

Even describing the self as a process is a metaphor with its own limits. Leonardo da Vinci was comparing poetry and painting when he made the remark:

Non sai tu che la nostra anima è composta di armonia?
[Do you not know that our soul is composed of harmony?]

He regarded painting as the higher art because by painting one can describe 'harmony' instantaneously so that it is immediately perceived by just 'watching' the painting, while writing poems, and music, need a step-like logic that one needs to 'hear' in sequence.

But I am already going further than language easily permits. Once you have climbed this particular ladder of understanding, you will appreciate that such matters are your own choice. We can choose our metaphors, they don't need to be imposed on us. It is time for me to leave my readers to their own thoughts.

The curtain call for this little book is that the artist disappears.

Bibliography

Anway, M.D., Cupp, A.S., Uzumcu, M., and Skinner, M.K. (2005). Epigenetic transgenerational actions of endocrine disruptors and male fertility. *Science*, **308**, 1466–9.

Armstrong, D.M. (1961). *Perception and the physical world*. London, Routledge & Kegan Paul.

Batchelor, S. (1994). *The awakening of the West: the encounter of Buddhism and Western culture*. Berkeley, Parallax Press.

Bateson, P. (2004). The active role of behaviour in evolution. *Biology and Philosophy*, **19**, 283–98.

Bennett, M.R. and Hacker, P.M.S. (2003). *Philosophical foundations of neuroscience*. Oxford, Blackwell.

Black, D.L. (2000). Protein diversity from alternative splicing: a challenge for bioinformatics and post-genome biology. *Cell*, **103**, 367–70.

Celotto, A.M. and Graveley, B.R. (2001). Alternative splicing of the *Drosophila Dscam* pre-mRNA is both temporally and spatially regulated. *Genetics*, **159**, 599–608.

Coen, E. (1999). *The art of genes: how organisms make themselves*. Oxford University Press.

Colvis, C.M., Pollock, J.D., Goodman, R.H., Impey, S., Dunn, J., Mandel, G. *et al.* (2005). Epigenetic mechanisms and gene networks in the nervous system. *Journal of Neuroscience*, **25**, 10375–89.

Cornman, J.W. (1975). *Perception, common sense and science*. New Haven and London, Yale University Press.

Crampin, E.J., Halstead, M., Hunter, P.J., Nielsen, P., Noble, D., Smith, N., and Tawhai, M. (2004). Computational physiology and the Physiome Project. *Experimental Physiology*, **89**(1), 1–26.

Crick, F.H.C. (1994). *The astonishing hypothesis: the scientific search for the soul*. London, Simon & Schuster.

Dawkins, R. (1976) *The selfish gene*. Oxford University Press.

Dawkins, R. (1976). Hierarchical organisation: a candidate principle for ethology. In *Growing points in ethology: based on a conference sponsored by St. John's College and King's College, Cambridge* (ed. P.P.G. Bateson and R.A. Hinde), pp. 7–54. Cambridge University Press.

Dawkins, R. (1982). *The extended phenotype: the gene as the unit of selection*. London, Freeman.

Dawkins, R. (2003). *A devil's chaplain.* London, Weidenfeld & Nicolson.

Deisseroth, K., Mermelstein, P.G., Xia, H., and Tsien, R.W. (2003). Signaling from synapse to nucleus: the logic behind the mechanisms. *Current Opinion in Neurobiology,* 13, 354–65.

Dover, G. (2000). *Dear Mr Darwin: letters on the evolution of life and human nature.* London, Weidenfeld & Nicolson.

Downer, L. (2003). *Madame Sadayakko: the geisha who seduced the West.* London, Headline.

Feytmans, E., Noble, D., and Peitsch, M. (2005). Genome size and numbers of biological functions. *Transactions on Computational Systems Biology,* 1, 44–9.

Foster, R. and Kreitzman, L. (2004). *Rhythms of life: the biological clocks that control the daily lives of every living thing.* London, Profile Books.

Gould, S.J. (2002). *The structure of evolutionary theory.* Cambridge, MA, Belknap Press of Harvard University Press.

Hardin, P.E., Hall, J.C., and Rosbash, M. (1990). Feedback of the *Drosophila* period gene product on circadian cycling of its messenger RNA levels. *Nature,* 343, 536–40.

Houshmand, Z., Livingston, R.B., and Wallace, B.A. (eds.) (1999). *Consciousness at the crossroads: conversations with the Dalai Lama on brain science and Buddhism.* New York, Snow Lion Publications.

Hunter, P.J., Robbins, P., and Noble, D. (2002). The IUPS Human Physiome Project. *Pflügers Archiv – European Journal of Physiology,* 445, 1–9.

International_HapMap_Consortium (2005). A haplotype map of the human genome. *Nature,* 437, 1299–319.

Jacob, F. (1970). *La Logique du vivant, une histoire de l'hérédité.* Paris, Gallimard.

Jablonka, E. and Lamb, M. (2005). *Evolution in four dimensions: genetic, epigenetic, behavioral, and symbolic variation in the history of life.* Cambridge, MA, and London, MIT Press.

Konopka, R.J. and Benzer, S. (1971). Clock mutants of *Drosophila melanogaster. Proceedings of the National Academy of Sciences,* 68, 2112–16.

Kövecses, Z. (2002). *Metaphor: a practical introduction.* Oxford University Press.

Kupiec, J.-J. and Sonigo, P. (2000). *Ni Dieu ni gene.* Paris, Seuil.

Lakoff, G. and Johnson, M. (2003). *Metaphors we live by.* University of Chicago Press.

Lamarck, J.-B. (1994). *Philosophie zoologique*; original edition of 1809 with introduction by Andre Pichot. Paris, Flammarion.

Levi, J. (2003). *Propos intempestifs sur le Tchouang-tseu.* Paris, Éditions Allia.

McMillen, I.C. and Robinson, J.S. (2005). Developmental origins of the metabolic syndrome. *Physiological Reviews,* 85, 577–633.

Maynard Smith, J. (1998). *Evolutionary genetics.* New York, Oxford University Press.

Maynard Smith, J. and Szathmáry, E. (1999). *The origins of life: from the birth of life to the origin of language.* New York, Oxford University Press.

Mayr, E. (1982). *The growth of biological thought: diversity, evolution and inheritance.* Cambridge, MA, and London, Belknap Press.

Monod, J. and Jacob, F. (1961). *Cold Spring Harbor Symposia Quantitative Biology.* **26**, 389–401.

Noble, D. (2002). The rise of computational biology. *Nature Reviews. Molecular Cell Biology*, **3**, 460–3.

Noble, D. and Noble, S.J. (1984). A model of sino-atrial node electrical activity based on a modification of the DiFrancesco-Noble (1984) equations. *Proceedings of the Royal Society of London, Series B*, **222**, 295–304.

Noble, D., Denyer, J.C., Brown, H.F., and DiFrancesco, D. (1992). Reciprocal role of the inward currents $i_{b,Na}$ and i_f in controlling and stabilizing pacemaker frequency of rabbit sino-atrial node cells. *Proceedings of the Royal Society of London, Series B*, **250**, 199–207.

Novartis_Foundation (1998). *The limits of reductionism in biology.* Chichester, Wiley.

Novartis_Foundation (2001). *Complexity in biological information processing.* Chichester, Wiley.

Novartis_Foundation (2002). *In silico simulation of biological processes.* London, Wiley.

Pichot, A. (1999). *Histoire de la notion de gène.* Paris, Flammarion.

Schrödinger, E. (1944). *What is life? The physical aspect of the living cell.* Cambridge University Press.

Smith, N.P., Pullan, A.J., and Hunter, P.J. (2001). An anatomically based model of transient coronary blood flow in the heart. *SIAM Journal of Applied Mathematics*, **62**(3), 990–1018.

Stelling, J., Klamt, S., Bettenbrock, K., Schuster, S., and Gilles, E.D. (2002). Metabolic network structure determines key aspects of functionality and regulation. *Nature*, **420**, 190–3.

Stevens, C. and Hunter, P.J. (2003). Sarcomere length changes in a model of the pig heart. *Progress in Biophysics and Molecular Biology*, **82**, 229–41.

Tomlinson, K.A., Hunter, P.J., and Pullan, A.J. (2002). A finite element method for an eikonal equation model of myocardial excitation wavefront propagation. *SIAM Journal of Applied Mathematics*, **63**, 324–50.

Wada, S. (2002). *The oxherder.* New York, George Braziller.

Watson, F.L., Puttnam-Holgado, R., Thomas, F., Lamar, D.L., Hughes, M., Kondo, M. *et al.* (2005). Extensive diversity of Ig-superfamily proteins in the immune system of insects. *Science*, **309**, 1874–8.

Index